Cambridge Elements ≡

Elements in the Philosophy of Science
edited by
Jacob Stegenga
University of Cambridge

CAUSATION

Luke Fenton-Glynn
University College London

CAMBRIDGE
UNIVERSITY PRESS

CAMBRIDGE
UNIVERSITY PRESS

University Printing House, Cambridge CB2 8BS, United Kingdom

One Liberty Plaza, 20th Floor, New York, NY 10006, USA

477 Williamstown Road, Port Melbourne, VIC 3207, Australia

314–321, 3rd Floor, Plot 3, Splendor Forum, Jasola District Centre, New Delhi – 110025, India

103 Penang Road, #05–06/07, Visioncrest Commercial, Singapore 238467

Cambridge University Press is part of the University of Cambridge.

It furthers the University's mission by disseminating knowledge in the pursuit of education, learning, and research at the highest international levels of excellence.

www.cambridge.org
Information on this title: www.cambridge.org/9781108706636
DOI: 10.1017/9781108588300

First published 2021

A catalogue record for this publication is available from the British Library.

ISBN 978-1-108-70663-6 Paperback
ISSN 2517-7273 (online)
ISSN 2517-7265 (print)

Causation

Elements in the Philosophy of Science

DOI: 10.1017/9781108588300
First published online: June 2021

Luke Fenton-Glynn
University College London
Author for correspondence: Luke Fenton-Glynn, l.glynn@ucl.ac.uk

Abstract: This Element provides an accessible introduction to the contemporary philosophy of causation. It introduces the reader to central concepts and distinctions (type vs token causation, probabilistic vs deterministic causation, difference-making, interventions, overdetermination, pre-emption) and to key tools (structural equations, graphs, probabilistic causal models) drawn upon in the contemporary debate. The aim is to fuel the reader's interest in causation, and to equip them with the resources to contribute to the debate themselves. The discussion is historically informed and outward-looking. 'Historically informed' in that concise accounts of key historical contributions to the understanding of causation set the stage for an examination of the latest research. 'Outward-looking' in that illustrations are provided of how the philosophy of causation relates to issues in the sciences, law, and elsewhere. The aim is to show why the study of causation is of critical importance, besides being fascinating in its own right.

Keywords: causation, counterfactual theories of causation, probabilistic causation, regularity theories of causation, causal modelling

ISBNs: 9781108706636 (PB), 9781108588300 (OC)
ISSNs: 2517-7273 (online), 2517-7265 (print)

Contents

1 Introduction

We humans take a great interest in causation. Causal knowledge helps us to understand, predict, and influence the world around us. A baby quickly comes to realise that pushing a button on her toy causes it to play a song; an adult exploits her knowledge of the effects of chamomile to sooth the baby's teething gums. Because of the close relation between causation, understanding, prediction, and control, natural and social scientists devote significant time and resources to investigating causal questions. Among other things, they ask or have asked: What are the causes of cancer? Of climate change? Of anomalies in the orbit of Uranus? What caused the extinction of the dinosaurs? The First World War? The 2007-8 financial crisis? Trump's election? The Covid-19 outbreak? What causes mental health problems? Crime? Price inflation?

Causation is of importance to psychology: if we want to understand how humans learn and reason, we need to understand their capacity for *causal* learning and reasoning. It's of importance in AI: if we want computers and robots to learn as well as (or better than!) humans, and to manipulate the world as (or more!) effectively, we need to programme them to be able to acquire causal knowledge and to use it.

Causation is also closely tied to questions of moral and legal responsibility. In a landmark UK legal case, the wife of the late Arthur Fairchild successfully sued Glenhaven Funeral Services[1] over her husband's death from mesothelioma – a type of lung cancer caused by asbestos exposure. In order to establish the company's responsibility for Fairchild's fatal illness, it was of course necessary to establish that there was a causal link between something they'd done – namely negligently expose Fairchild to elevated levels of asbestos – and the illness itself.

Despite its ubiquity and importance, it's surprisingly difficult to say exactly what causation is. Difficult questions about the fundamental nature of the world – especially those that don't readily admit of empirical resolution – naturally attract the attention of philosophers. But causation isn't only of intrinsic philosophical interest. Greater theoretical clarity on its nature has had significant payoffs in the sciences and in law. And, close to home for philosophers, it has payoffs in virtue of the fact that causation plays a role in key theories of a variety of philosophically interesting phenomena including (but not limited to) reference, perception, decision, knowledge, inference, action, and explanation.

It shouldn't be thought that work on the theory of causation is the exclusive preserve of philosophers. Much important theoretical work has been done by

---

[1] *Fairchild v Glenhaven Funeral Services Ltd* ([2002] UKHL 22; [2003] 1 AC 32).

computer scientists, economists, statisticians, legal scholars, psychologists and others – reflecting the broad, interdisciplinary importance of a better understanding of causation. Fortunately, these days, there's significant interaction between theorists in these various disciplines, which has enriched our collective understanding of causation. A prime example of the payoffs of this cross-disciplinary interaction is the theory of causal modelling that we'll examine as part of Sections 4 and 5.

This Element examines some of the progress that's been made in understanding the nature of causation as well as some of the unresolved challenges. Since the literature on causation is large, this Element is, of necessity, a selective introduction. It focuses on three broad traditions within the theory of causation: the regularity, counterfactual, and probabilistic approaches. Perhaps the most contentious omissions are the process approach – which seeks to analyse causation in terms of causal processes with the latter understood, on the most promising such account, as the world-lines of objects that possess conserved quantities (Dowe 1992, 2000; Salmon 1994, 1997) – and the New Mechanist approach – exemplified by Machamer et al. (2000) and Glennan (2017), among many others. I've made this choice, not because I don't think that understanding causal processes and mechanisms is of vital importance (I do!), but because I don't think that these are causal bedrock: I think there are relations of causation that are more fundamental than the notion of a causal process or mechanism and that an adequate understanding of processes and mechanisms will require an adequate understanding of these more fundamental causal relations.[2] Regularity, counterfactual, and probabilistic approaches are attempts to understand these fundamental causal relations.

This view is contentious as process theorists and some New Mechanists think that the fundamental causal relation(s) *can* be understood in terms of processes or mechanisms. For example, Dowe (2000, 90) seeks to define causal interactions in terms of processes, while Glennan (1996) suggests that causation – at least outside the domain of microphysics – might be analysed in terms of mechanisms. However, process theories have been plagued by the problem of distinguishing genuine causal processes from world-lines that don't correspond to processes without falling back on an appeal to some more basic causal relation, perhaps understood in terms of counterfactual dependence (see, e.g., Hitchcock 1995, 2009; Choi 2002).

As regards mechanistic approaches, there's certainly no consensus among New Mechanists that the notion of mechanism is prior to that of causation. As

---

[2] The notion that there might be more than one fundamental causal relation is taken up in Section 3.4.2.

Craver and Tabery (2019) note, 'Mechanists have disagreed with one another about how to understand the cause in causal mechanism. ... Four ways of unpacking [it] have been discussed: conserved quantity accounts, mechanistic accounts, activities accounts, and counterfactual accounts'.

I've already said that it's doubtful that the conserved quantity approach can yield an understanding of the fundamental causal relation(s). The activities approach, on the other hand, is a primitivist approach (see Craver and Tabery 2019), whereas we'll be examining accounts that seek a deeper understanding of causation. Meanwhile, the counterfactual approach is one that we'll be exploring in Section 4. Finally, the mechanistic account – as advocated by Glennan (1996) – is regressive. The proposal is that causal connections that may seem basic at (say) the biological level can be understood in terms of mechanisms at the chemical level, and those at the chemical level in terms of mechanisms at the physical level. There's thus a hierarchy of mechanisms. The concern, though, is that this hierarchy bottoms out at the level of fundamental physics at which level we have causings that can't be mechanistically understood. Again, this favours the view that there are fundamental causal relations in terms of which mechanisms can ultimately be understood. The regularity, counterfactual, and probabilistic approaches seem the most promising approaches to understanding these basic causal relations.

2 Regularity Theories of Causation
2.1 Hume

Though Western theorising about causation dates back at least to Aristotle (*Physics* 195 a 4–14; *Metaphysics* V.2), David Hume (1739, 1748) is rightly considered the father of the modern tradition of attempts to understand that relation. Hume is standardly interpreted as advocating a *regularity theory* of causation.

Specifically, according to Hume (1739 I.iii.2), causes occur temporally prior to their effects, and are either contiguous with them in space and time or else connected to them by a contiguous 'chain' of causation. For example, a person may break a window by throwing a rock at it despite the fact that the throw occurs a short interval of time before and a few metres away from the breaking because, once the rock is thrown, it traces a continuous trajectory until it hits and shatters the window. The window's breaking is caused by the throw via this 'chain' (there's no 'action at a distance'), with the position and momentum of the rock at each stage on its trajectory being caused by its previous positions and momenta and by the throw itself, and with the window's shattering being caused by the prior states of the rock all the way back until we get to the throw itself.

But we don't have a case of causation just any time an event occurs prior to and contiguously with another. Towards the end of the movie *Saving Private Ryan*, in a defiant last stand, Captain Miller repeatedly fires his pistol at a German Tiger tank. The bullets are of course completely incapable of piercing the tank's armour. Down to his last bullet, Miller points his gun and shoots at the tank at the very moment the tank is blown to pieces by a bomb dropped by a US P-51 aircraft. The impact of Miller's bullet is immediately prior to, and contiguous with, the explosion of the tank. Yet it's the bomb and not the bullet that causes the tank to explode.

Fortunately, Hume's account doesn't imply that the bullet impact was a cause. That's because, in addition to priority and contiguity, Hume adds a third requirement: *constant conjunction*. For Hume, for an event c to be a cause of an event e, it must be the case that events like c are always followed contiguously by events like e.[3] This criterion excludes the impact of Miller's bullet from counting as a cause of the tank's explosion. That's because events like the former aren't always followed by events like the latter. Indeed, Miller had already fired his gun at the tank five times prior to firing his last bullet: the impact of none of these previous five bullets was followed contiguously by an explosion. So Hume's analysis yields the correct verdict about this case.

Although Hume doesn't say this, it's tempting to think that not any old constant conjunction can ground a causal relation, but rather one might wish to require that the constant conjunction be entailed by the laws of nature. This avoids problems such as the following. Suppose there exists an extremely rare isotope, call it 'unobtanium-352'. Only one atom of this isotope ever exists. Suppose this atom happened to decay on the afternoon of November 19, 1863, immediately before Lincoln delivered the Gettysburg Address and contiguously with it. Now, for any type T of event of which the Gettysburg Address is a member ('famous speeches', say), it's true that all cases of unobtanium-352 decay are followed by events of type T. Nevertheless, it clearly doesn't follow that the decay of the unobtanium-352 atom was a cause of the Gettysburg Address.[4] A sophisticated regularity theory can avoid this

---

[3] We could, on Hume's behalf, distinguish *direct* from *indirect* causation, with c counting as a direct cause of e iff c is prior to, and contiguous with, e and events like c are always followed contiguously by events like e. *Indirect* causation would then be understood in terms of chains of (i.e. ordered sequences of events that stand in relations of) direct causation.

[4] Note that even if we take 'constant conjunction' to require a multiple instances, we're still liable to get 'accidental' constant conjunctions that aren't apt to underwrite causal relations (see Armstrong 1983, 15–17).

conclusion by pointing out that the fact that all instances of unobtanium-352 decay are followed by events of type T isn't entailed by the laws of nature (rather, it's an instance of an accidental regularity).[5]

2.2 Mill

An apparent problem with Hume's account is that sometimes we seem to have causation without constant conjunction. John Stuart Mill pointed this out in making the following observation:

> It is seldom, if ever, between a consequent and a single antecedent that ... invariable sequence subsists. It is usually between a consequent and the sum of several antecedents ... . In such cases it is very common to single out only one of the antecedents under the denomination of Cause, calling the others merely Conditions. Thus, if a person eats of a particular dish, and dies in consequence ... people would be apt to say that eating of that dish was the cause of death. There needs not, however, be any invariable connexion between eating of the dish and death; but there certainly is, among the circumstances which took place, some combination or other on which death is invariably consequent: as, for instance, the act of eating of the dish, combined with a particular bodily constitution ... . The real Cause, is the whole of these antecedents ... . (Mill 1843, III.v.3)

Mill's idea, then, is that constant conjunction ('invariable sequence') rarely obtains between a single earlier event ('antecedent') and a single later event ('consequent'). Taking Mill's example, it could easily happen that someone dies from eating particular dish (because, say, it contains an allergen) but that others who eat it survive. Mill thinks that those who die have features that distinguish them from those who don't. For instance, they may have a severe allergy to an allergen in the dish. In this case, the constant conjunction

---

[5] A couple of comments are worth making regarding this proposed appeal to laws of nature. *First*, whether one regards it as marking a departure from a pure regularity theory of causation will depend upon one's preferred metaphysics of laws. It will not be a departure if one adopts a *regularity theory of laws*. Whilst sophisticated regularity theories of laws – such as the Best System Analysis (Lewis 1994) – regard laws as regularities, they don't count just any old regularity as a law of nature. Thus, for instance, it's to be hoped that they wouldn't count an unobtainium-325 decay/famous speech regularity as a law.

Second, for the appeal to laws of nature to be satisfactory, it will presumably be necessary that a wide range of regularities outside the domain of fundamental physics (e.g. 'aspirin consumption is followed by pain relief') are entailed by laws of nature. That's because it's clear that our causal claims extend to such domains. There's an extensive philosophical literature on the status of generalisations outside of fundamental physics. For overviews of important aspects of this literature, see Cat (2017) and Reutlinger et al. (2019). Thanks to an anonymous referee for encouraging me to say something about both of the foregoing points.

is between eating the dish *and (e.g.) having a severe peanut allergy* (as Mill puts it: having 'a particular bodily constitution') and death.

Mill thinks that, properly speaking, in such a case it's the combination of the severe peanut allergy with the eating of the dish that's the cause of death. However he observes elsewhere (Mill 1843, III.v.3) that, in ordinary talk, we often single out just one of the factors in such a combination as the cause and regard the others as mere 'conditions'. Specifically, he claims that we're inclined to pick out '*events*' or '*changes*' like the eating of the dish as causes and treat long-standing '*states*' like the possessing of the peanut allergy as mere '*conditions*'.

2.3　Hart and Honoré on the Cause/Condition Distinction

As noted in the Introduction, the study of causation is a truly interdisciplinary endeavour and the profound contributions of the legal scholars H. L. A. Hart and Tony Honoré to our understanding of it is the perfect illustration of this. One of their contributions was to further investigate the relationship between those factors that we pick out as 'causes' and those that we label mere 'conditions'. According to them, this distinction is one 'to which common sense adheres in face of the demonstration that cause and conditions are "equally necessary" if the effect is to follow' (Hart and Honoré 1959, 33). In giving an account of the grounds on which we make such distinctions, they emphasise the close connection between causation and explanation.

Hart and Honoré (1959, 35) point out that the sorts of effect that tend to pique our interest – and lead us to seek causes to explain – are abnormal events. In Mill's example, this was the sudden death of a person. Previous examples that we've used include the shattering of a window and the explosion of a tank. Hart and Honoré consider the example of the outbreak of a fire. They point out that, in order to explain an abnormal event, some abnormal factor typically needs to be cited. For example, in order to explain why a building was destroyed by fire, emphasising the presence of oxygen or flammable material wouldn't typically be that helpful, whereas pointing out that a lit cigarette was dropped would be much more so. The reason is that oxygen and flammable material were *always* present, but what we need for an explanation is to know what *made the difference* between this occasion, on which the building burned down, and all the previous times when it didn't (Hart and Honoré 1959, 35). It's the abnormal factor, the dropping of the cigarette, that's the difference-maker and hence we're liable to call it the 'cause' while treating the others as 'mere conditions' (Hart and Honoré 1959, 35).

Normal factors will often be what Mill described as 'states' – ongoing or permanent factors – while abnormal factors will often be 'changes' or 'events'. For instance, the presence of oxygen and flammable material are typically relatively permanent states, while the dropping of a cigarette is a 'change'. But Hart and Honoré show that things aren't that simple. They give the following example:

> If a fire breaks out in a laboratory or in a factory, where special precautions are taken to exclude oxygen during part of an experiment or manufacturing process, since the success of this depends on safety from fire, there would be no absurdity at all in *such* a case in saying that the presence of oxygen was the cause of the fire. The exclusion of oxygen in such a case, and not its presence, is part of the normal functioning of the laboratory or factory ... .
> (Hart and Honoré 1959, 35)

Suppose that the manufacturing process involves the frequent production of sparks. Then, it seems, we might pick out the presence of oxygen as a cause even if the leak occurred some time before a spark was produced and the fire started. If that's right, then we would appear to have a case where the relatively longstanding state (the presence of oxygen) is picked out as the cause, while the change (the spark) might be treated as a mere condition. This would suggest that the distinction we draw between causes and conditions tracks the abnormal/normal distinction rather than the event/state distinction where these two come apart.

2.4 Mackie

As we saw, although Mill acknowledges that in ordinary talk we distinguish between causes and conditions, he thinks that the 'real Cause' is the combination of all the factors needed to bring about the effect. We also saw that Hart and Honoré seek to account for why we distinguish between causes and conditions in terms of our explanatory interests, noting that these various factors may be equally necessary for the effect. Especially if we think that the event/state distinction does not always track the cause/condition distinction (as Hart and Honoré's example suggests), we might conclude that the cause/condition distinction isn't really a metaphysical distinction at all but rather one that's to be accounted for by a suitable pragmatics of causal talk. It's tempting to impute this view to David Lewis when he says:

> We sometimes single out one among all the causes of some event and call it 'the' cause, as if there were no others. Or we single out a few as the 'causes,' calling the rest mere 'causal factors' or 'causal conditions.' ... We may select the abnormal or extraordinary causes, or those under human control, or those

we deem good or bad, or just those we want to talk about. I have nothing to say about these principles of invidious discrimination. I am concerned with the prior question of what it is to be one of the causes (unselectively speaking). (Lewis 1973a, 558–9)

We'll return to Lewis's account of causation in Section 4. For now it's worth noting that, even if we agree that there's no deep metaphysical distinction between causes and conditions (or indeed between the factor that we might on some occasion be inclined to pick out as 'the' cause of some effect, and the rest of that effect's causes), we needn't follow Mill in taking the conjunction of all the factors requisite to produce an effect as the real cause. An alternative approach – as suggested in the passage from Lewis – is that we allow that each of the factors in question counts as *a* cause, so that a given effect may have a plurality of causes.

The latter is the approach of Mackie, whose well-known account of causation perhaps marked the zenith of the regularity approach.[6] Mackie's account can be introduced by means of his own example:

> Suppose that a fire has broken out in a certain house .... Experts ... conclude that it was caused by an electrical short-circuit at a certain place. ... Clearly the experts are not saying that the short-circuit was a necessary condition for this house's catching fire at this time; they know perfectly well that ... the overturning of a lighted oil stove, or any one of a number of other things might, if it had occurred, have set the house on fire. Equally, they are not saying that the short-circuit was a sufficient condition for this house's catching fire; for if the short-circuit had occurred, but there had been no inflammable material nearby, the fire would not have broken out .... In what sense, then, is it said to have caused the fire? At least part of the answer is that there is a set of conditions ... including the presence of inflammable material, the absence of a suitably placed sprinkler ... which combined with the short-circuit constituted a complex condition that was sufficient for the house's catching fire – sufficient, but not necessary, for the fire could have started in other ways. Also, of *this* complex condition, the short-circuit was an indispensable part: the other parts of this condition, conjoined with one another in the absence of the short-circuit, would not have produced the fire. ... In this case, then, the so-called cause is ... an *insufficient* but *necessary* part of a condition which is itself *unnecessary* but *sufficient* for the result. ... [L]et us call such a condition (from the initial letters of the words italicized above), an inus condition. (Mackie 1965, 245)

---

[6] This isn't to denigrate subsequent accounts within the regularity tradition, including the excellent contributions of Strevens (2007) and Baumgartner (2013). But these can be regarded as attempts to revive the tradition after a prolonged period in which it has been out of favour. It's fair to say that it currently remains a minority approach.

Clearly Mackie's account makes heavy use of the notions of *necessary* and *sufficient* conditions. The reason it's usually classed as a regularity theory is that necessary and sufficient conditions can themselves be understood in terms of regularities.[7] For instance, we might say that the set of conditions comprising the short-circuit, the presence of inflammable material, the presence of oxygen, the absence of a sprinkler, etc. is *sufficient* for the fire iff whenever such a constellation of factors co-occurs, a fire always ensues. Likewise, we might say that the short-circuit is a *necessary* (or non-redundant) element of this set of conditions iff it's not the case that whenever the remaining conditions in the set co-occur then fire ensues.[8]

With the notions of 'necessity' and 'sufficiency' so interpreted, we can view Hume as taking causes to be sufficient for their effects (and possibly necessary too, if 'constant conjunction' is taken to cut both ways, so that not only does a cause never occur without its associated effect, but an effect never occurs without its associated cause), while Mill agrees that real causes are sufficient for their effects but takes them to typically be complex (e.g. the complex condition comprising the short-circuit, the presence of oxygen, the presence of inflammable material, and the absence of a sprinkler system, etc.). Mackie, on the other hand, allows non-redundant elements of such sets (e.g. the short-circuit) to count as causes.

2.5 Problems with the Regularity Theory

Perhaps the most important problems that have been raised against the regularity approach – and which have contributed to its decline in popularity in recent years – are (i) the problem of the direction of causation; (ii) the problem of probabilistic causes; (iii) the problem of common causes. We'll see in later sections that other theories of causation aren't immune to these problems, but they're often thought to be particularly intractable for the regularity approach.

2.5.1 The Direction of Causation

Causation is almost always supposed to be an asymmetric relation: if *a* causes *b*, then *b* doesn't also cause *a*. One might object that feedback loops provide a counterexample: depression causes someone to drink, but alcohol itself acts as a depressant. The rejoinder is that, while a person's depressed state at time t_1

---

[7] Though, drawing upon the point made in Section 2.1, we may wish to understand them in terms of *lawfully entailed* regularities.

[8] Actually, Mackie himself proposes that necessity and sufficiency only indirectly be understood in terms of regularities: specifically, he proposes to interpret necessity and sufficiency in terms of counterfactuals, with the counterfactuals understood in terms of regularities (Mackie 1965, 253–5). So one might take Mackie's account to be a sort of hybrid regularity/counterfactual theory. Counterfactual theories will be examined in Section 4.

might cause her to drink at time t_2, which in turn causes her depressed state at time t_3, what we'd never have is a person's depressed state at t_1 causing her to drink at t_2 with her drinking at t_2 causing her depression at t_1. The point is that, where a and b are particular events or states that obtain at specific times, it can't be the case both that a causes b and that b causes a. This point is sometimes put by saying that *token causation* (that is, the causal relation between token – i.e. particular, dated – events or states) is asymmetric. We'll have more to say about token causation – and its distinction from what's sometimes known as *type* causation – in Section 3.3.

One might wonder whether it's really entirely impossible that a should be a token cause of b and b a token cause of a. The General Theory of Relativity (GTR) allows for the possibility of what are known as 'closed time-like curves' – paths in space-time of positive distance that lead from a given space-time point p back to p that could be traversed without ever travelling at or above the speed of light. Suppose object o traverses such a path from p to p and goes through some point q on its way. Then the state of o at p (its position and velocity) is presumably a cause of its state at q and its state at q a cause of its state at p. But, whilst GTR allows that this is a possibility, we don't have reason to think that closed time-like curves in fact exist in our universe.

The trouble for regularity theorists is that they face a challenge in accounting for why we don't have bi-directional causation even in quoditian cases. That's because necessity and sufficiency are two sides of the same coin: if a is sufficient for b then b is necessary for a. Putting it in regularity-theoretic terms, if events like a are always accompanied by events like b, then there are no a-like events without b-like events. This means that a challenge arises for the accounts of Hume and Mill when we have causes that are both necessary and sufficient for their effects. It also turns out that often when a is an inus condition of b, then b is also an inus condition of a.

This difficulty isn't necessarily fatal to the regularity theory. What it shows is that some extra element needs to be added to the analysis to distinguish cause from effect. Hume takes this extra element to be *time*: he requires that the cause be earlier in time than the effect. One potential objection to Hume's approach is that it's not clear how to reconcile it with the possibility of closed causal loops of the sort seemingly allowed by GTR. Mackie was reluctant to rule out backwards-in-time causation, and so instead took the cause to be the event that becomes *fixed* first (Mackie 1980, ch. 7). While, ordinarily, earlier events become fixed before later ones (because becoming a past event is the usual way in which an event becomes fixed), Mackie thought that in special circumstances (of the sort that would allow for retro-causation) a later event might become fixed prior to an earlier one. Yet it's not clear that Mackie's proposal allows the sort of bi-directional causation that appears to be compatible with GTR, since it's at best unclear in what sense, in such a case, each of the causal relata could become fixed prior to the other.

But, even if the problem of distinguishing causes from effects can be overcome, there are other troubling objections to the regularity theory.

2.5.2 Probabilistic Causes

Consider the following example, given by Elizabeth Anscombe (1971, 24)[9]:

Probabilistic Bomb

"[A] bomb is connected with a Geiger counter, so that it will go off if the Geiger counter registers a certain reading; whether it will or not is not determined, for it is so placed near some radioactive material that it may or may not register that reading. … [T]here would be no doubt of the cause of the reading or of the explosion if the bomb did go off."

According to quantum mechanics (at least on orthodox interpretations), the decay of radioactive isotopes is an irreducibly probabilistic process. For instance, there's no feature that distinguishes those tritium atoms that decay within 12.32 years (the half-life of tritium) from those that don't. Each tritium atom simply has a fundamental chance of 0.5 of doing so. Thus if we place a chunk of tritium next to the Geiger counter described in **Probabilistic Bomb**, it's not determined by the state of the tritium when it's placed, nor even this together with the state of the entire universe at that time, whether sufficient atoms will decay so that enough β particles are emitted within a short enough time that the Geiger threshold is reached and the bomb explodes. Rather, there's simply a certain probability that this will happen.

Nevertheless, as Anscombe notes, it's clear that if the Geiger threshold is reached and the bomb explodes, then the placing of the tritium near the Geiger is a cause. Correspondingly, the person who placed the material there – perhaps together with the person who set the whole contraption up – would presumably be held legally and morally responsible for the explosion and its consequences. This poses an insurmountable problem for the regularity analyst for, despite being a cause, the placing of the tritium isn't sufficient, even when taken together with any or all of the circumstances that obtain at the time it's placed, for the bomb's explosion. The best the regularity theorist can do is restrict the scope of the analysis, offering it as an analysis of *deterministic*

---

[9] A similar example is given by the physicist Richard Feynman (1965, 147).

causation, and allowing that probabilistic causation must be analysed in other terms. We'll examine accounts of probabilistic causation in Section 5.

2.5.3 Common Causes

Perhaps the most serious problem of all for regularity analyses – and the one that prompted Mackie himself (Mackie 1980, ch. 3) to abandon a pure regularity theory of causation[10] – is that of common causes. A well-known example is given by Hans Reichenbach (1971, 193): namely of a fall in atmospheric pressure which results in (i) a certain barometer predicting a storm; and (ii) a storm.

The structure of this example is illustrated in Figure 1, with a representing the fall in atmospheric pressure, b the barometer's prediction of a storm, and s the storm itself, and with the 'arrows' representing causal relations.[11] Event a is a 'common cause' of events b and s.

This sort of structure poses problems for the regularity approach. Suppose the fall in atmospheric pressure is sufficient for the storm, or at least there's a set of conditions X in combination with which it's sufficient for the storm. Suppose also that the barometer is in perfect working order so that it will only indicate a storm if there's been a fall in atmospheric pressure. Then the barometer's prediction is an inus condition of the storm: the set of conditions comprising the barometer reading, the barometer being in perfect working order, and X, is

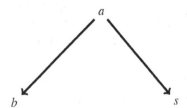

Figure 1 Common Causation

---

[10] Likewise, the problem is one that Strevens (2007) seeks to deal with via an account that, while incorporating elements of a Mackie-inspired regularity analysis, also appeals to a notion of 'causal connection'. Though Strevens (2007, 110) suggests that this notion of 'causal connection' might potentially be understood via a 'process theoretic' account of causation (of the sort briefly discussed in the Introduction of this Element) or a nuanced counterfactual approach, he refrains from endorsing any specific analysis of the notion.

[11] This representation isn't to be confused with the graphs of causal models that will be introduced in Sections 4 and 5. Rather, it's merely an intuitive representation of the causal relations that obtain in this case.

sufficient for the storm,[12] while the latter two of these conditions aren't on their own sufficient. Requiring that causes temporally precede their effects doesn't help since barometers are prediction devices and therefore, if the barometer is any good, it will signal a storm prior to the storm's occurrence.

In the face of these difficulties, the regularity approach has fallen out of favour. In Sections 4 and 5 we'll explore alternatives to it. But first we'll pause to discuss some important general questions about the nature of causation and our methodology in investigating it. It'll be easier to do this now we've embarked on our discussion of theories of causation, since this will lend greater concreteness to the discussion.

3 Interlude: Some Important Background

Before examining alternatives to the regularity approach, it's worth making explicit some points that have hitherto been in the background, and making some further observations and distinctions regarding causation. Many of the issues discussed in this section are more general than the question of whether this or that particular theory of causation is best, but will also augment our understanding of the theories discussed in this Element.

3.1 Is Causation Really a Relation? What Are Its Relata?

So far, we've spoken as though causation is a *relation* and that's indeed the view taken by almost everyone who has theorised about it. Thus, for instance, if the short-circuit caused the fire, then presumably the short-circuit stands in a causal relation to the fire. On the assumption that causation is a relation, we may discuss the formal properties of this relation. For instance, we've already briefly discussed whether it's asymmetric – whether (necessarily) for all x and y, if x causes y, then it's not the case that y causes x – or whether it's simply that bidirectional causation is something unfamiliar to our everyday experience. In Section 4, we'll see that there's also a question of whether causation is *transitive*: whether (necessarily) for all x, y, and z, if x causes y and y causes z, then x causes z. A related question is whether causation is irreflexive – whether (necessarily) for all x, it's the case that x doesn't cause x. This is related because if causation isn't asymmetric (i.e. there's some x and y such that x causes y and y causes x) but is transitive, then it follows that it's not irreflexive

---

[12] This means that Mill's account yields an incorrect result about the causation of the storm. Hume's account, on the other hand, gives incorrect results whenever we have a common cause c that's necessary for an effect e_1 and sufficient for an independent effect e_2, for this will mean that e_1 is sufficient for c and therefore e_2.

(from the fact that x causes y and y causes x, we can infer by transitivity that x causes x).

Also on the assumption that causation is a relation, we might ask what the *relata* of the causal relation are. The standard answer is *events*, where the latter category is broadly construed to include what might be considered 'states' or 'conditions' (such as *the presence of oxygen*). Thus, for instance, Davidson (1967) and Kim (1973a) advance considerations in favour of the view that causation relates events, on a reasonably broad construal of what counts as an 'event'.

There's a hitch, though, with the view that causation is a relation with events as its relata. That hitch is how to make sense of statements like 'the crop failed because there was no rain', or 'the absence of rain caused the crop to fail', which seems a legitimate paraphrase. The difficulty is that it's not clear that an absence of an event is itself an event. Moreover, we can't just happily acknowledge that and simply allow that causation is a relation that may have events *or absences* as its relata. The reason is that it's not obvious that absences are particulars that could stand in such a relation to events or to other absences (Beebee 2004, cf. Mellor 1995, 163–5).

What are our options? One is to deny that causation is a relation at all. This option is pursued by Mellor (1995). Mellor (1995, 156) regards the basic form of causal truth as 'E because C', where 'C' and 'E' are facts, with facts understood as true sentences (Mellor 1995, 161). Facts aren't particulars and so aren't apt to be the relata of a real relation. Moreover, in cases like 'the crop failed because there was no rain', Mellor (1995, 163–5) claims there are no particulars that serve as truthmakers for C and E that could themselves stand in a real relation of causation.

A second option, discussed by Lewis (1986b, 190–3, 2004b, 281–2), would be to insist that absences *are* particulars and therefore apt to stand in a causal relation. Lewis discusses two versions of this view: (i) there are negative events (events whose essence is the absence of something); (ii) there are (positive) events that are contingently absences. To illustrate the idea behind (ii), *turning right* at a junction might be one way for a car to fail to turn left; another way for it to fail to turn left might be by its *continuing straight on*. If the car in fact turns right, then it's turning right might be taken to be the absence of its turning left, but only contingently because it's possible that another event (the car's continuing straight on) should have instead been the absence of its turning left.

The trouble with (i) is that negative events would be strange sorts of things, and it's not obvious how to reconcile them with standard accounts of the ontology of events (e.g. Kim 1973a, Lewis 1986a). The trouble with (ii) is that there don't always seem to be events of the right sort to contingently serve as absences. For instance, what positive event could contingently serve as the absence of rain? The presence of sun? But sun tends to be beneficial to crops (so its presence, unlike the absence of rain, wouldn't explain a crop failure) and, in any case, its presence isn't incompatible with rain (fortunately for rainbow fans). The presence of a cloudless sky? But this is really an *absence* of clouds.

A third option (see, e.g., Beebee 2004) is to maintain that causation is a relation, but admit that absences aren't particulars that can stand in such a relation, and so deny that statements like 'the crop failed because there was no rain' have a causal relation as their truthmaker. If one nevertheless wishes to maintain that such statements can be true, then one takes on the burden of giving a semantics for them that doesn't appeal to the existence of a corresponding causal relation. Lewis (2004b, 282–3), for instance, appeals to the obtaining of certain counterfactuals. For instance, the statement 'the absence of rain caused the crop to fail' is made true by the truth of the counterfactual 'if there had been rain, then the crop wouldn't have failed' (which, Lewis would claim, doesn't itself require a causal relation as its truthmaker – Lewis 1979). A regularity theorist, on the other hand, might take this statement to be made true by the regularity that, whenever there's healthy crop growth (in the absence of a sprinkler system, etc.), it's preceded by rain. On the other hand, an advocate of a probabilistic analysis (of the sort to be discussed in Section 5) might take it to be made true by the fact that rain raises the probability of healthy crop growth.

In what follows, I'll adopt the standard view that causation is a relation and remain neutral between the second and third option for dealing with absences. But, setting aside the problem of absences, are there particulars other than events (construed liberally) that could serve as the relata of a causal relation? One might think that *objects* could. For instance, if a brick is thrown at a window, which shatters, it sounds perfectly acceptable to say 'the brick caused the window to break'. Yet the brick is an object, not an event.

It is, however, tempting to think that event causation is fundamental. That's because, whenever it sounds felicitous to say 'object o was a cause of e', it seems there's some event c that o participated in such that it's apt to say that 'o was a cause of e in virtue of o's participation in c' and also that 'c was a cause of e'. The brick caused the window to break, but it did so *by* hitting the window, and the brick hitting the window is an event which itself caused the window to break. On the other hand, whenever it sounds felicitous to say 'a was a cause of object o' (e.g. 'the collision of the Indo-Australian and Eurasian plates caused the Himalayas'), an apt paraphrase seems to be 'a was a cause of o's coming into existence', and the *coming into existence of o* is an event.

Likewise, we commonly speak of *agents* as causes (agents might be regarded as a certain sort of object). For example, one might felicitously say 'Aisha broke the window'. But again it seems that Aisha broke the window by participating in an event – the throwing of the brick – that caused the window to break. Some, however, have argued that there are instances of agent causation that are irreducible to event causation (see, e.g., O'Connor 2002). Often such arguments are motivated by attempts to account for freedom of the will: the thought being that irreducible agent causation allows agents to originate causal chains, whereas if agents caused things merely in virtue of participating in events, then they wouldn't exhibit genuinely free action, since those events would themselves be answerable to a chain of prior causes. But the view that there's irreducible agent causation is highly controversial, and won't be pursued here.

Finally, some take *variables* or their *values* to be (among) the relata of the causal relation.[13] There's a certain ambiguity in the notion of a 'variable'. On the one hand, it's sometimes used to refer to a particular piece of mathematical formalism. In this sense, 'X' can be used as a variable, with 'x' as one of its values. On the other hand, 'variable' can be used to refer to the sort of thing representable by the mathematical formalism. In this sense, the speed of a car is a variable, with 90mph being one of its possible values. It's the sort of thing represented by the formalism rather than the formalism itself that's taken by some to be a potential causal relatum.

Where variable *values* are taken to be the relata[14] this view seems compatible with the view that the causal relata are events (when 'event' is construed liberally, to include states/conditions). After all, such things as the car's travelling at 90mph can be seen as events (in this liberal sense). This view doesn't escape the issues surrounding absences, however. For instance, we might take the amount of rain that falls in some location during a particular crop season to be representable by a variable R, with $R = 0$ corresponding to no rain and positive values of R corresponding, say, to the total rainfall in millimetres. We might also represent the success of a certain crop with a variable C that, for example, takes $C = 0$ if the crop dies, but with positive values of C corresponding, say, to the height in centimetres that the crop achieves by harvest. In that case, adopting the language of variables, it would be perfectly reasonable

---

[13] This view is particularly associated with the *causal modelling* approach to causation to be explored in sections 4.6 and 5.3–5.6. Pearl (2009), Spirtes et al. (2000), and Halpern (2016) are prime examples of this approach.

[14] As we'll see in Section 4.6.5, taking the variables themselves to be the relata might sometimes be appropriate depending on the variety of causation at issue.

– if, in fact, $R = 0$ and $C = 0$ – to say that $R = 0$ was a cause of $C = 0$. We're then confronted again with the question of exactly what makes such a claim true.

3.2 How Many Relata?

Assuming causation is a relation, we might ask how many relata it has. The answer that most naturally springs to mind is 'two'. When one says 'c is a cause of e', '... is a cause of ...' is a two-place predicate which might naturally be taken to pick out a binary relation (specifically, a relation that obtains between pairs of events) in the world. Yet not everyone takes causation to be binary. For instance, Hitchcock (1996) argues that it's a ternary relation with the relata being three events: two actual and one merely possible. The idea is that causal relations are best described by claims of the form 'c *rather than c′* is a cause of e'. For instance, consider the interrogative sentence: *Did the car's travelling at 60mph cause the crash?* Arguably this is unanswerable because the question asked by this sentence is underdetermined. By contrast, consider the pair of sentences: *Did the car's travelling at 60mph rather than 40mph cause the crash?* and *Did the car's travelling at 60mph rather than 80mph cause the crash?* These seem to ask more determinate questions and are more tractable. Arguably, then, the problem with the first sentence is that it asks *did c cause e simpliciter*, but there are no facts about whether c caused e simpliciter, only facts about whether c *rather than c′* caused e for various possible alternatives, $c′$ to c. The second and third sentences ask determinate questions because they make clear what causal fact is at issue.

Others (e.g. Schaffer 2005, 2013; Northcott 2008) go further and argue that causation is a *quaternary relation* which finds expression in sentences of the form 'c *rather than c′* caused e *rather than e′*'. To illustrate, suppose Mo can take no dose, one dose, or two doses of medicine and that doing so will lead to no recovery, slow recovery, or quick recovery, respectively. Then claims of the following form seem entirely cogent: 'taking one dose rather than two doses caused Mo to recover slowly rather than quickly'. More generally, it has been suggested (Schaffer 2005, 327–9; Northcott 2008) that we should take the contrasts to c and e to be (possibly singleton) *sets* of alternatives. For instance, it seems sensible to ask whether the car's travelling at 60mph rather than *less than 60mph* caused the crash, but of course the contrast here appears to concern a *set* of specific alternatives to the car's travelling 60mph.

It is of course true that we often make claims of the form 'c caused e' with no contrasts to c or e specified. Those who take causation to be contrastive on the cause and/or effect side respond that such sentences manage to express propositions because *context* supplies the contrasts. For instance, if an assertion such as "drinking lots of wine last night caused my hangover" is in most contexts evaluable as true without explicitly specified contrasts, that's because in most contexts the implicit contrast to my drinking all that wine is some sort of default behaviour like my not drinking alcohol (and the implicit contrast to my having a hangover is just my not having a hangover). But perhaps in rare contexts the implicit contrast might be different so that the sentence expresses a false proposition. For example, suppose that last night I was participating in an initiation ceremony where I was forced to choose to drink either lots of wine or lots of whisky. Then the implicit contrast to my drinking lots of wine is my drinking lots of whisky, so that the sentence can be considered elliptical for "drinking lots of wine rather than lots of whisky last night caused my hangover (rather than my not having a hangover)", which is presumably false.

3.3 Token vs Type

A distinction can be drawn between token (or 'singular') causal claims and generic (or 'type') causal claims. Examples of generic causal claims include *smoking causes cancer* or *quantitative easing causes house price rises*. Examples of token causal claims include *Jane's smoking caused her cancer* or *the Bank of England's 2009–12 quantitative easing programme caused the London house price bubble that began in 2012*. Token causal claims concern individual events (such as *Jane's smoking* and *her developing cancer*), while generic causal claims concern types of event (such as *smoking* and *cancer*).

Scientists are interested in assessing the truth both of token causal claims – *the Chicxulub impact caused the K-Pg Extinction event, the Big Bang caused the Cosmic Microwave Background, the financial crisis of 2007–8 caused the Great Recession* – and generic causal claims – *smoking causes cancer, loose monetary policy causes inflation, abnormal protein folding is a cause of allergies*. Lawyers are principally interested in the truth-values of token causal claims (*this act caused that harm*).

A fairly common view among philosophers is that type causal claims are some sort of generalisation over token causal claims. As Lewis (1973a, 558) puts it: 'Presumably [type causal claims] are quantified statements involving causation among particular events (or non-events)'. This view yields a parsimonious metaphysics of causation, since it implies that there's no *sui generis* relation of type causation in addition to token causation. Still, as Lewis (1973a, 558) admits, type causal claims don't admit of a straightforward analysis in terms of the standard quantifiers of first order logic. One plausible suggestion is that the quantifier involved is a generic operator (Carroll 1991, Eagle 2015).[15]

But the metaphysical priority of token causation isn't a completely uncontroversial matter, and what view one takes on this issue is likely to be in part informed by what one thinks is the best theoretical approach to causation. For instance, on the regularity approach, token causation is understood in terms of associations between event types. But then, for the regularity theorist, it seems natural simply to interpret such associations as type-level causal relations. By contrast, as we'll see in Section 4, it's natural for advocates of counterfactual theories of causation to take token causation to be the more basic notion.[16]

Although the type vs token distinction is one of the most commonly drawn and important when it comes to varieties of causal claim, it's not exhaustive of those that can be distinguished. For instance, Woodward (2003) defines a number of different causal relations that might obtain between *variables*. As we'll see in Section 4.6.5 and Section 5.6, it's doubtful that *any* sort of causal relation between variables falls neatly into the 'token' or 'type' categories, and indeed there may be a case for treating such relations as *sui generis*, even if not entirely unrelated to token and type causation.

3.4 Aims and Method

When engaged in almost any task, it's helpful to ask *what, exactly, are we trying to achieve?* An answer to this will help us to frame an appropriate method for reaching our goals. The task of theorising about causation is no different.

There are a variety of sometimes subtly distinct aims that one might have in theorising about causation. For instance, one might be engaged in a reductive or a non-reductive project, one might be engaged in something like 'conceptual analysis', or (to the extent that this differs) one might be concerned with causation as it is 'out there in the world'. If engaged in conceptual analysis, then one might be concerned with analysing the layperson's concept of causation or (insofar as these differ) the scientific or legal concept of causation. Also, one might be concerned with the project of 'descriptive' analysis – examining the contours of the concept as it actually is – or prescriptive/revisionary analysis – seeking to outline a more useful and/or more coherent concept of causation. If investigating causation 'in the world', one might be concerned with what causation contingently is in our world (Dowe 2000, ch. 1), or with what it is as

---

[15] Approaches to the semantics of the generic operator are discussed by, for example, the contributors to Carlson and Pelletier (1995).

[16] And, as we'll see in Section 5.6, the question of the relation between token and type causation is somewhat complex if one adopts a probabilistic approach.

a matter of metaphysical necessity (i.e. across all possible worlds). Examining these various sorts of project, and situating the accounts of causation examined in this Element within them is the task for this section.[17]

3.4.1 Concepts vs Metaphysics

Hume famously distinguished between 'our idea of causation' and causation 'in the objects' (in the world) (Hume 1739, I.iii.14). It's tempting to regard investigation of the nature of the former as corresponding to the project of conceptual analysis, and investigation of the latter as a metaphysical project. Unfortunately, things aren't quite so straightforward for reasons that will be described at the end of this section.

Hume's own primary focus was on the *idea* of causation. The extent of the implications he took his discussion to have for the nature of causation 'in the objects' is a matter of some controversy (see Read 2014). Hume was particularly interested in the *aetiology* of our idea of causation. This was part of a more general project of showing that all human ideas (concepts) ultimately derive from experience. Hume acknowledged that causation was a tricky case. Central to our idea of causation, he thought, is the idea that there is a *necessary connection* between cause and effect. Yet Hume claimed that when we observe interactions that we take to be causal – for example, of one billiard ball impacting upon, and imparting motion in, a second – we don't observe any such necessary connection (Hume 1739, I.iii.13). Ultimately, Hume traces our idea of necessary connection to a certain introspective experience: our alleged experience, upon observing a particular event, of feeling 'a determination of the mind' (i.e. an involuntary compulsion) to expect its usual effect(s).

The example of Hume makes it clear why an investigation of the *idea* of causation isn't automatically, or in any direct way, an examination of the phenomenon as it exists in the world. For, although necessary connection is (according to Hume) a key element of our concept of causation, 'necessity is something, that exists in the mind, not in the objects' (Hume 1739, I.iii.14), and when we speak of necessary connections in the world, the word 'necessity' is '*wrongly apply'd*' (Hume 1739, I.iii.1; italics original). For Hume, causation in the objects can be captured simply by defining a cause as '[a]n object [viz. event] precedent and contiguous to another, and where all the objects resembling the former are plac'd in like relations of precedency and contiguity to

---

[17] Of necessity, the discussion here is brief. Readers are referred to Paul and Hall (2013, esp. ch. 2, sect. 3) for a more extensive discussion of the goals and methods of a range of contemporary approaches to caussation.

those objects, that resemble the latter' (Hume 1739, I.iii.14). This, of course, is Hume's regularity account of causation.[18]

Hume, as well as other British empiricists (notably John Locke), were important influences in the later analytic tradition which gave the method of conceptual analysis a central role in philosophy. We've already seen that there is some reason to think that analysis of the concept of causation isn't automatically an investigation of causation as it exists in the world. Still, the two projects aren't entirely unrelated. If one gave an account of causation in the world that bore little resemblance to our concept of causation, then one could reasonably be accused of changing the subject. For example, this would be a fair charge if one asserted that causation 'in the objects' were identical to mere correlation. After all, it's a common aphorism that correlation doesn't imply causation: barometer readings don't affect the weather, for instance.

There's a certain amount of give and take here, however. Perhaps nothing 'in the objects' *could* perfectly answer to our concept of causation. This would be so if, for instance, there are no necessary connections in nature, but Hume is right that the concept of causation involves the idea of 'necessary connection' as a key component. One response would be to simply deny that there is such a thing as causation 'in the world'. Yet if there is something in the world that answers near enough to our concept of causation, or if there's something in the world that our concept of causation tracks with at least a reasonable degree of systematicity (for instance, Hume thought that we were systematically willing to apply our concept of causation in cases in which there's a confluence of priority, contiguity, and constant conjunction), then it may be reasonable to call this thing 'causation'.

If one regards causation 'in the world' as diverging from our concept of causation, then it might be natural to suggest a revision to our ordinary concept so that it better corresponds to causation in the world or so that it better corresponds to something that it would be useful for our concept of causation to track. This is one reason why one might endorse a 'revisionary' or 'prescriptive' account of the concept of causation, as opposed to a purely descriptive one. There are other (perhaps related) reasons for endorsing a revisionary analysis. For example, one might think that the ordinary concept of causation is somehow *imprecise* or even *incoherent* and that we would be better served by a concept that was coherent and precise. We'll see in Sections 4 and 5 that

---

[18] It's a matter of controversy whether Hume thought that this 'definition' captures all that causation is in the world or merely all that we can know about causation as it is in the world – see (Read 2014). The former thesis seems fairly clearly the one argued for in (Hume 1739). There is, however, some evidence that he rows back from this in (Hume 1748).

especially some of the more formal accounts of causation (e.g. those appealing to structural or probabilistic models) arguably involve a degree of precision that our ordinary concept of causation lacks.[19] Indeed, as we'll see, some of those who have developed such accounts (e.g. Spirtes et al. 2000, Hitchcock 2001b, Woodward 2003, Pearl 2009) define overtly technical causal notions – for example, 'total', 'contributing', 'direct' causation – which, though arguably useful, are sometimes only loosely related to the causal concept(s) of the layperson. Nevertheless, several of these authors also discuss the notion of what they call 'actual causation' (Woodward 2003, sect. 2.7; Pearl 2009, ch. 10), which is taken to be more-or-less the ordinary person's token causal concept.

When theorising about causation, it can often be helpful to have in mind the question of *what the point is* in our having a concept of causation, or perhaps a concept of this or that sort of causation (e.g. a concept of type causation and a concept of token or actual causation). This is perhaps particularly obvious if one is in the business of developing a prescriptive analysis of causation – since a consideration of what practical or cognitive needs are served by a concept (or concepts) of causation will help guide the development of a concept that better meets those needs. But it is also the case even when it comes to developing a descriptive analysis: an understanding of *why* we have a concept (or concepts) of cause can help us to more successfully delineate the contours of the concept(s) that we *do* have, just as an understanding of the purpose or function of a carburetor can help us to understand (and perhaps improve) its design.

Mellor (1995, 58–66, ch. 7) argues that causation has several important 'connotations', including that effects are evidence for their causes, that causes explain their effects, and that causes are means for bringing about their effects as ends. The connections to evidence, explanation, and practical reasoning can well explain *why* we have a concept or concepts of causation and should inform our theorising about it. For instance, if one were to devise a theory of causation on which it were difficult or impossible to see why – on such a theory – causes are evidence for, explanatory of, or means to, their effects, then this would be at least prima facie evidence that we were barking up the wrong tree. It turns out that this is a genuine issue for many accounts of causation. For instance, each of the broad traditions that we consider in this Element – the regularity,

---

[19] This is also true of some of the less technical accounts. For instance, we'll see that Lewis's counterfactual analysis of causation implies that overdeterminers are not causes in cases of so-called symmetric overdetermination. Lewis (1986b, 208) believes that there is a 'lack of firm common-sense judgements' regarding whether overdeterminers are genuine causes and says that when, as in cases of overdetermination, 'common sense falls into indecision or controversy … then … [w]e can reasonably accept as true whatever answer comes from the analysis [of causation] that does best on the clearer cases'.

counterfactual, and probabilistic traditions – has, at its core, a simple insight: that causes are constantly conjoined with their effects, or that effects (counter-factually) depend upon their causes, or that causes raise the probability of their effects. These basic insights tend to be relatively conducive to seeing how our concept of causation serves its purpose: if a cause is constantly conjoined with its effect, then it's good evidence for that effect (and vice versa); if an effect depends upon its cause, then the cause appears to be a means (or at least a part of a means) to its effect; if a cause raised the probability of its effect, then it would seem that it helps explain why the effect happened, etc.

The trouble is that these 'insights' aren't quite right. As we've seen, not all causes are constantly conjoined with their effects (e.g. probabilistic causes). We'll also see that not all effects counterfactually depend upon their causes, nor do all causes raise the probability of their effects. Thus credible philo-sophical accounts of causation are more subtle than simply equating causation with constant conjunction, counterfactual dependence, or probability-raising. Yet this greater complexity tends to come associated with a greater difficulty of explaining *why* these accounts capture a notion that serves the functions that we suppose our concept(s) of causation to have. Whilst we may be able to see, roughly at least, why an event c upon which another event e depends is a means to e, or why c can help explain e if c is a probability-raiser of e, it's rather more difficult to see why c should count as a means to e if e merely bears to c the *transitive closure* of the counterfactual dependence relationship (see Section 4.1) or if c is merely a de facto probability-raiser of e (Section 5.4). It therefore seems that if one is to properly justify such an account, one incurs the burden of showing how it captures a concept that does the work our concept of causation does. This is an issue taken up in Section 5.7.

A final point worth making before leaving the topic of conceptual and met-aphysical analysis is that there is a certain amount of blurring of the lines in contemporary philosophy. These days when a philosopher offers an 'analy-sis' of a concept like 'causation' or 'knowledge' it's rare for the analysans to be couched in terms that would be particularly easy for the layperson to understand. For example, popular contemporary analyses of *knowledge* invoke technically-defined notions of 'safety' and/or 'sensitivity'. Analyses of causation invoke such things as the transitive closure of the relation of non-backtracking counterfactual dependence. It's not particularly plausible that such analyses are articulating the structure implicit in the ordinary lan-guage user's concept of causation. Rather, they're engaged in something like the project of 'simply' providing truth-conditions for causal claims. Argua-bly at least, it's a short step from the latter project to making claims about the metaphysical truthmakers for causal claims. Insofar as this characterisation of the contemporary project of 'conceptual analysis' is correct, criticisms of such analyses as *obscurum per obscurius* may *sometimes* be unfair. They're not intended to be anything like dictionary definitions, but are perhaps better regarded as preludes to metaphysical accounts. Critiques may therefore be best directed at the claims of the extensional adequacy of the truth-conditions and/or the attractiveness of the corresponding metaphysical picture of causation.[20]

The fact that Hume's investigation of the 'idea of causation' is an attempt to articulate the structure of the ordinary concept is the reason I hesitated previously to equate it with the modern project of conceptual analysis. There's also a reason for caution about equating his project of investigating 'causation in the objects' with metaphysical analysis. This is that, when Hume defines causation 'in the objects', it's not entirely clear whether he means to be giving an account of what causation objectively is *as a matter of metaphysical necessity* (i.e. across all possible worlds) or what it objectively is *contingently* (i.e. in our world, and perhaps in worlds that share some important feature with our world – such as having the same laws of nature).

Most of the major traditions in the contemporary philosophy of causation seek the former sort of account. But not all do. For instance, Dowe (2000, ch. 1) explicitly advances his account of causation – according to which causation consists of the transference of conserved quantities – as what he calls an 'empirical analysis': an analysis of what causation contingently is in the actual world. He doesn't wish to deny that there might be causation in worlds in which there are no conserved quantities, or in which there are no conserved quantities of the sort required to underwrite causal relations.

The fact that many philosophers focus on what might be called 'metaphysical analysis' rather than 'empirical analysis' can help to explain aspects of the structure of the debate. For example, we have seen that regularity accounts of causation have been criticised for being unable to handle probabilistic causation. It also turns out that some probabilistic accounts of causation are criticised for not being able to handle deterministic causation! If both deterministic and probabilistic causation are genuine possibilities, then an account that can handle only one or the other won't make for a satisfactory metaphysical analysis. But, for instance, if it turns out (contrary to the prevailing assumption) that all

---

[20] Thus, for instance, though the notion of 'non-backtracking counterfactual dependence' may be obscure as far as the ordinary competent language user is concerned, if this relation is grounded in some asymmetry of physics – as Lewis (1979) and Dunn (2011) have argued – then giving extensionally adequate truth-conditions for causal claims in terms of non-backtracking counterfactual dependence potentially paves the way for an overall simplification of our metaphysics.

causation in our world is deterministic then a regularity analysis might still be advanced as an empirical analysis.[21]

3.4.2 Causal Proliferation?

It has already been noted that ordinary language users make both type and token causal claims. So prima facie it looks like we'll need to make room for the existence of at least two causal notions (albeit it may turn out that one is conceptually or metaphysically reducible to the other). It's plausible that the notions of type and of token causation serve slightly different uses. If we recall Mellor's 'connotations' of causation, the notion of type causation seems particularly useful when it comes to prediction or means–ends reasoning, while token causation is closely tied to explanation. Thus, for instance, to explain the COVID-19 outbreak we investigate its token causes. Part of the reason this is important is that presumably there are features of whatever the token causes were that are generalisable. Thus, for instance, if one of the token causes turns out to have been an instance of bat–livestock interaction, this might help to confirm a type-causal claim about the risks of such interactions which will hopefully help us to reduce human virus outbreaks in future.

In any case, it looks as though we have, and have use for, at least two causal notions. But it's also an interesting question whether there might be, or be grounds for, still further proliferation of causal concepts. Regarding actual causation, Pearl (2009, 309) points out that '[h]uman intuition is extremely keen in detecting and ascertaining this type of causation'. To the extent that intuitions about type causation are less firm, it might be that there's no canonical variety of type-causation that's picked out by our ordinary thought and talk. This perhaps gives us the flexibility to define various type-causal notions as and when they are useful without doing violence to causal intuition and without being susceptible to accusations of changing the subject. In this context, it's interesting to note that when Woodward (2003, sect. 2.3) characterises a variety of technical causal notions – 'direct', 'total', and 'contributing' cause – he regards them each as varieties of type causation, and distinguishes them from actual causation. As we'll see in Sections 4 and 5, these technical notions are useful in analysing relations between variables in complex causal systems.

---

[21] In fact, it might be that deterministic and probabilistic causation coexist in some worlds (perhaps at different scientific 'levels'), and that ours is such a world. For relevant discussion, see Albert (2000), Loewer (2001), Ismael (2009), Werndl (2011, 2263), Emery (2015), and List and Pivato (2015, sect. 9). Potentially, then, there's an interesting discussion to be had about the legitimacy of seeking a philosophical account of causation that applies only to certain domains (e.g. levels) *even within a world.*

And perhaps one can also usefully define technical causal notions at the token-level provided that one is careful to distinguish these from our notion of actual causation.[22]

But what would be particularly interesting, given our firm intuitions about actual causation, is if there were some inconsistency in our thought or talk about it that might motivate a distinction between two or more concepts of actual causation. There are at least two ways in which this might occur. First, it might be that the conception of actual causation in play varies across contexts. For instance, it might be that lawyers or scientists or historians draw upon a different conception of actual causation from one another or from ordinary folk. Second, it might be that there's an inconsistency in our thought or talk about actual causation even within a context.

The latter is a view argued for in some detail by Hall (2004) who thinks that the ordinary folk concept of actual causation bifurcates into a concept of 'dependence' and a concept of 'production'. In many cases, production and dependence align. For example, a short circuit might produce a fire, with the occurrence of the latter also depending upon the occurrence of the former. Sometimes, though, they come apart: if two short circuits occur simultaneously, each sufficient for the fire, then whilst both may play a role in the fire's production, the occurrence of the fire might depend on neither (since in the absence of each, the other short circuit would have still been enough to produce the fire). This is an instance of so-called 'symmetric overdetermination' (a phenomenon that will be discussed in more detail in Section 4.4). To the extent that we feel conflicted about whether each of the short circuits was a cause in such a case, the breaking of the normal association of production and dependence might be one explanation for this feeling of conflict.

There are also prima facie reasons for thinking that the conception of actual causation varies across contexts. For instance, a simple counterfactual 'but for' test for actual causation is preponderant in the law ("But for the short circuit, would the fire have occurred?"). In apparently stark contrast, there's considerable controversy about the legitimacy of using counterfactual reasoning to establish causal claims in history (see, e.g., Evans 2013). Still, this prima facie divergence may *only* be that. When application of the 'but for' test produces counterintuitive consequences, judges tend to vary the test for causation.[23]

---

[22] For instance, Hitchcock (2007, 503–4) introduces a technically defined notion of 'token causal structure' which he clearly distinguishes from the ordinary notion of actual causation. And, as we'll see in Section 4.6.5, Woodwardian notions of 'direct', 'total', and 'contributing' cause can also be defined for the token level.

[23] Examples include *R v Dyson* [1908 2 KB 454 (CA)], *Kingston v Chicago & N.W. Ry* [191 Wis. 610, 211 N.W. 913 (1927)], *McGhee v National Coal Board* [1972 3 All E.R. 1008, 1 W.L.R. 1], and *Fairchild v Glenhaven Funeral Services Ltd* [2002 UKHL 22].

Indeed, the so-called 'plain meaning rule' of statutory interpretation[24] dictates that lawyers ought not to depart from the ordinary language meaning of terms (including causal terms) employed in statute provided that these are precise and unambiguous. On the other hand, despite the fact that many historians overtly repudiate counterfactuals, it has been pointed out that "[h]istorians ... use counterfactuals regularly though implicitly when they assign necessary causes, and sometimes in assigning degrees of importance to causes" (Tucker 1999, 265). It's therefore not clear that there's any hard evidence that either lawyers or historians operate with a concept of actual causation that's distinct from the folk concept.

Surveying the usage of token causal concepts across the sciences is clearly far too big an undertaking to pursue here, but there are some prima facie reasons to be skeptical that scientific usage differs dramatically from usage in ordinary life. One reason is that much of our ordinary reasoning about causes is clearly proto-scientific. When I try to figure out what caused the tripping of a circuit breaker in my house, my investigation seems to be at least broadly continuous with scientific investigation of token causes. Indeed, Hitchcock and Knobe (2009, 591) provide an interesting conjecture about the purpose of the ordinary notion of actual causation: namely, 'the concept of actual causation enables us to pick out appropriate targets for intervention'. By this they have in mind that the ordinary concept tends to pick out as causes those factors that could have been manipulated to avoid or to influence the effect. (In this instance, preventing my toddler from tipping her apple juice onto a power block would have done the trick!) If this is right, then not only does it connect actual causation – which we previously noted seems closely tied to the 'explanatory' connotation of causation – with the connotation concerning control, it suggests that our ordinary concept of actual causation serves a purpose that is also useful to scientists, which would make continuity between the ordinary and scientific concepts unsurprising.

3.4.3 Reduction

Reductivity is often seen as a virtue of philosophical analyses. For those seeking a metaphysical reduction of causation, success would – other things being equal – allow for a simplification of their overall metaphysic. For example, if causation could be reduced to constant conjunction with the latter interpreted

---

[24] This 'rule' is a convention followed by the courts of England and Wales, with similar conventions also adopted in other jurisdictions. It was articulated in the *Sussex Peerage Case* [1844; 11 Cl and Fin 85] as follows: 'If the words of the Statute are in themselves precise and unambiguous, then no more can be necessary than to expound those words in that natural and ordinary sense.'

simply in terms of tokens of event types standing in regular spatio-temporal relations of priority and contiguity, then – since it is likely that spatio-temporal relations, as well as event types and tokens, will feature elsewhere in our metaphysical theory anyway (and not just in our account of causation) – our overall metaphysic will be simplified since causation will not need to feature as a basic posit of it.

To the extent that a conceptual analysis seeks to elucidate the truth-conditions for causal statements as a prelude to our investigation of what the metaphysical truth-makers might be, we might hope that those truth-conditions can be given without invoking causal notions, for otherwise it seems that our resulting metaphysical account may end up being non-reductive. On the other hand, if, like Hume, we're seeking a 'conceptual analysis' that articulates the structure of the ordinary person's concept of causation, we might favour a reductive account to the extent that – like Hume – we think there's a prima facie mystery about that concept that might be dispelled by analysing it into non-causal constituents.

Yet in spite of the fact that many philosophers of causation have pursued reduction, some (e.g. Woodward 2003, 104–7) have persuasively argued that analyses of causation can be informative even if they are non-reductive: that is, even if the analysans isn't entirely free of causal notions. To see how this might work, consider, by way of analogy, the following two biconditionals:

1. S knows p iff S knows p.
2. S knows p iff S knows that one knows that p.

To assert the first biconditional would be to be guilty of the most blatant unilluminating circularity. By contrast, whilst the second biconditional obviously couldn't serve as a reductive analysis of knowledge (since the verb 'to know' appears on both sides of the biconditional), it nevertheless makes an interesting and substantive assertion about knowledge.[25] For instance, it says something about knowledge that (if true) clearly distinguishes it from propositional attitudes like hope, regret, etc.[26]

The regularity theories examined in Section 1 appear to fall into the category of reductive analyses: causation is analysed in terms of (lawful) regularity, which itself doesn't appear to be an inherently causal notion. Later in this Element, however, we'll consider accounts (such as the 'interventionist' approach

---

[25] This biconditional is implied by the (controversial) KK principle together with the factivity of knowledge.

[26] I don't claim that knowledge is itself a propositional attitude, though this is a view defended by Williamson (2000, ch. 1).

discussed in Section 4.6.6) that are non-reductive in nature. In evaluating these accounts it is important to have in mind the question of whether they are illuminating even if they aren't reductive.

4 Counterfactual Theories of Causation

Despite the fact that they're usually associated with the regularity approach to causation, both Mill and Hume suggest a close connection between causation and counterfactuals. Thus Mill says, '[I]f a person eats of a particular dish, and dies in consequence, that is, *would not have died if he had not eaten it*, people would be apt to say that eating of that dish was the cause of death' (Mill 1843, III.v.3; italics mine). Mill treats the italicised counterfactual as synonymous with the causal claim that the person died in consequence of eating the dish. Similarly, Hume at one point (Hume 1748, sect. VII) offers a counterfactual definition of causation. It has been pointed out (Lewis 1973a) that counterfactual definitions in fact aren't equivalent to standard regularity-theoretic definitions.

The counterfactual definition implicit in the aforementioned Mill passage is equivalent to the 'but for' test for causation in the law: a cause is *an event but for which the effect wouldn't have occurred*. The seminal philosophical work developing a counterfactual theory of causation is (Lewis 1973a). As we'll see in the next subsection, the counterfactual analysis developed there is more sophisticated than the 'but for' criterion.

Before proceeding, it's worth pointing out that counterfactual approaches standardly focus on token causation. This is natural, since it makes good sense to talk about the (actual and counterfactual) occurrence and non-occurrence of token events (e.g. *Jane's smoking* and *Jane's developing lung cancer*). By contrast, it's not clear that it makes sense to talk about the (actual and counterfactual) occurrence and non-occurrence of event types (e.g. *smoking* and *lung cancer*) unless this is simply understood in terms of the occurrence and non-occurrence of instances of those types, in which case we're back to talking about token events.

A second point to note before proceeding is that we'll assume determinism in this section. That is, we'll assume that given the state of the world at one time, the laws of nature determine how it will be at all future times. We'll see in Section 5 that the counterfactual approach can be generalised to cases where determinism doesn't hold. But things will be simpler if we focus on the deterministic case for now.

4.1 Lewis's 1973 Analysis of Causation

Let c and e be token events, and let $O(c)$ and $O(e)$ be the propositions c *occurs* and e *occurs*, respectively. Then, where $\square\!\!\rightarrow$ expresses the counterfactual connective 'If it had been that ... , then it would have been that ... ', Lewis (1973a, 562–3) says that e causally depends[27] upon c iff the following two counterfactuals hold:

(i) $O(c) \square\!\!\rightarrow O(e)$.

(ii) $\neg O(c) \square\!\!\rightarrow \neg O(e)$

The first of these expresses the counterfactual 'if it had been the case that c occurred, then it would have been the case that e occurred' (for short: 'if c had occurred, then e would have occurred'); the second expresses the counterfactual 'if it hadn't been the case that c occurred, then it wouldn't have been the case that e occurred' (for short: 'if c hadn't occurred, then e wouldn't have occurred').[28]

Lewis (1973a, 1973b, 1979) gives counterfactuals a *possible worlds* semantics. According to him, a counterfactual $A \square\!\!\rightarrow C$ is true iff either there are no worlds in which A (i.e. A is necessarily false) or there's some world in which $A\&C$ that's more similar ('closer') to our own, actual world (hereafter '@'), than is any world in which $A\&\neg C$. Lewis (1973b, 14–15) subscribes to a principle known as Strong Centering, according to which @ is more similar to itself than any other possible world is to it. Given his semantics for counterfactuals, this has the immediate implication that any counterfactual with a true antecedent has the same truth-value as its consequent. This means that, where c and e are actual events (as they must be if they're to be contenders to stand in a causal relation in the actual world), the aforementioned counterfactual (i) is automatically true. This means that whether e causally depends upon c hinges upon whether counterfactual (ii) is true.

Lewis, however, doesn't equate causation with causal dependence between events.[29] Rather, he takes this merely to be a sufficient condition for causation.

---

[27] Sometimes I'll say 'counterfactually depends' rather than 'causally depends' to express this relation in what follows. Lewis reserves 'counterfactually depends' for the relation between the *propositions* $O(e)$ and $O(c)$ that obtains when the following pair of counterfactuals are true. I'll be more lax in also using it to describe the relation between the corresponding events too.

[28] Following Lewis, I'm calling both subjunctive conditionals that have true antecedents and those with false antecedents 'counterfactuals'. Lewis gives a uniform semantics to both types of subjunctive.

[29] Strictly speaking, we should say 'distinct' events. To give Kim's (1973b, 571) example: '[i]f I had not written 'r' twice in succession, I would not have written "Larry"'. Yet writing 'r' twice in succession isn't a cause of writing 'Larry' but rather an essential part of it.

Instead, causation is equated with the transitive closure (or 'ancestral') of causal dependence. That is, c is a cause of e iff there's a sequence of actual events c, d_1, ..., d_n, e such that e causally depends upon d_n, which causally depends upon ... d_1, which causally depends upon c (Lewis 1973a, 563). Lewis (1973a, 563) calls such a sequence a 'causal chain'. The case where e causally depends upon c is the special case of a 'one link' causal chain. Lewis identifies causation with the ancestral of causal dependence because he takes causation to be transitive. Yet, as Lewis is aware, c can be connected to e via a causal chain without its being the case that e causally depends upon c. This happens in cases of what Lewis (1986b, Postscript E) calls *early preemption*.

Before discussing early preemption, it's worth noting that the question of whether causation is transitive is contested (see, e.g., Hall 2000, 2004, Lewis 2004a, Hitchcock 2001a, Sartorio 2005). There are apparent counterexamples. Suppose a terrorist plants a bomb on a train but, fortunately, the bomb is spotted and the train safely evacuated. The bomb's being spotted counterfactually depends upon and is caused by its being planted. The survival of all the passengers counterfactually depends upon and is caused by the evacuation. Yet it's at best dubious whether the bomb's being planted causes the passengers to survive. Lewis's account in terms of the ancestral of counterfactual dependence implies that it does. And Lewis can't readily drop his commitment to transitivity by identifying causation with counterfactual dependence rather than its ancestral because, as we'll see, this would deprive him of the ability to handle cases of early preemption.

4.2 Early Preemption

Cases of preemption are cases in which, in the absence of a cause of some particular effect, another event (the 'preempted alternative') would have brought about that effect instead. Lewis (1986b, Postscript E) distinguishes between 'early preemption' and 'late preemption'. In cases of *early* preemption, as Lewis (1986b, 200) puts it, 'the process running from the preempted alternative is cut off well before the main process running from the preempting cause has gone to completion'. By contrast, in cases of *late* preemption, this 'backup process' isn't cut off before the effect itself occurs.

The following is an example of early preemption based on one given by Edgington (1997, 420). An example of late preemption will be considered in the next subsection.

Early Preemption

Alice and Bob are hunting deer. They have one gun between them and are
down to their last bullet. They spot a deer. Both are ace shots and the shot
is a simple one. If Alice doesn't shoot, then Bob will. Alice shoots. The
bullet hits the deer and the deer dies.

In **Early Preemption**, although Alice's shot is a cause of the deer's death, the
deer's death doesn't counterfactually depend upon her shot: *but for* her shot,
Bob would have shot and so the deer would still have died. Bob's intention to
shoot if Alice doesn't can be considered the preempted backup. Had Alice not
shot, this would have initiated a process (involving Bob's shooting) that would
itself have resulted in the deer's death. However, such a process is stopped
in its tracks *before* the deer's death (which is why this counts as a case of
early preemption). Specifically, as soon as Alice shoots, Bob is prevented from
shooting.

Cases of preemption pose a problem for regularity theories. Alice's shooting
together with other prevailing conditions such as the shot being an easy one,
the gun's being in good working order, etc., suffices for the deer's death (recall
that, for now, we're assuming determinism). The trouble is that Bob's intention to shoot if Alice doesn't, together with his being an excellent shot, Alice's
being an excellent shot, the shot being an easy one, the gun's being in good
working order, etc., is *also* a sufficient condition (it guarantees that, one way
or the other – by means of Alice's shooting or Bob's shooting – the deer's death
will come about). Bob's intention is a non-redundant element of this sufficient
condition (after all, the condition doesn't include Alice's shooting). So Bob's
intention is an inus condition of the deer's death. Yet Bob's intention isn't a
cause.[30]

Returning to Lewis's account, observe that, despite the fact that Bob's intention means there's no counterfactual dependence of the deer's death on Alice's
shot, Lewis would claim there's a chain of counterfactual dependence connecting the two, and this is why Alice's shot counts as a cause of the deer's death.
Specifically, consider the event comprising the speeding bullet's presence at the
mid-point between Alice and the deer. Call this event *m*. Lewis's (1973a, 567,
1986b, 200–1) suggestion would be that there's a two-link chain of counterfactual dependence connecting Alice's shot and the deer's death, with the deer's

---

[30] Although he develops an account of causation that draws upon the notion of an inus condition,
Strevens (2007, esp. 5.4) recognises the need to supplement the notion of an inus condition with
a notion of 'causal production' (which he doesn't define in regularity-theoretic terms) in order
to deal with cases like this.

death counterfactually dependent upon *m*, and *m* counterfactually dependent upon Alice's shot. That one of these links holds is fairly obvious: it's intuitive that, if Alice hadn't shot, then *m* wouldn't have obtained (there might instead have been a speeding bullet mid-air between *Bob* and the deer). Lewis would also claim that, if *m* hadn't obtained then the deer wouldn't have died. Specifically, Lewis claims that the nearest worlds in which *m* doesn't occur are worlds in which Alice still shoots (but perhaps shoots off target) and so Bob doesn't shoot. We therefore shouldn't be tempted to think that had *m* not occurred, then that would have been because Alice didn't shoot, so Bob would have shot, and the deer would have died anyway. The latter sort of reasoning relies on what (Lewis 1979, 456) calls 'backtracking' counterfactuals (which assert that if the state of the world at some later time had been different, then its state at some earlier time would have been different – e.g. if the later event *m* hadn't occurred, then the earlier event of Alice's shooting wouldn't have occurred) and claims that a correct counterfactual semantics, at least for the purposes of evaluating causal claims, is one that rules backtracking counterfactuals false.

It's this prohibition on backtracking counterfactuals that's also Lewis's proffered solution to two of the key difficulties for the regularity theory examined in Section 2.5: namely, the problem of distinguishing cause from effect (considered in Section 2.5.1) and the problem of the spurious implication of causation between independent effects of common causes (discussed in Section 2.5.3). Lewis (1973a, 564–5) claims that the counterfactual analysis of causation gets the direction of causation right because effects counterfactually depend upon their causes (or counterfactually depend upon events that counterfactually depend upon … their causes), but (in virtue of non-backtracking) causes don't counterfactually depend upon their effects (or counterfactually depend upon events that counterfactually depend upon … their effects).

Now consider the example of common causation discussed in Section 2.5.3 in which the fall in atmospheric pressure is a cause both of the barometer reading and the storm. According to Lewis (1973a, 565–7), his counterfactual analysis avoids the spurious implication that the barometer reading is a cause of the storm because (in virtue of non-backtracking) it's false that if the barometer hadn't predicted a storm, the atmospheric pressure wouldn't earlier have fallen. Rather it would still have fallen and so the storm would still have occurred. So the storm neither counterfactually depends on the barometer reading nor is connected to the barometer reading by a chain of counterfactual dependence that goes via the earlier fall in atmospheric pressure.

Lewis (1979) provides a detailed story about why his possible worlds semantics for counterfactuals implies the 'non-backtracking' result.[31] His approach has, however, been criticised by several authors (Elga 2001; Woodward 2003, ch. 3.6; Kment 2006). A now-popular alternative approach to providing a non-backtracking semantics is the so-called 'interventionist' approach (Goldszmidt and Pearl 1992; Meek and Glymour 1994; Woodward 2003), which we'll examine in Section 4.6.6. But however we understand the semantics for counterfactuals, let's grant that Lewis's analysis correctly handles cases of early preemption (even if it does, rather controversially, commit to causal transitivity). Still the account runs into trouble in cases of late preemption, as Lewis (1986b, 203–5) himself admits.

4.3 Late Preemption

Edgington's example can be modified to give a case of late preemption.

Late Preemption
Alice and Bob are hunting deer. They each have a gun and each gun has plenty of ammunition. They spot a deer. If Alice doesn't shoot *or if Alice shoots but fails to kill the deer*, then Bob will shoot and kill the deer. Alice shoots and kills the deer.

It's worth emphasising how **Late Preemption** differs from **Early Preemption**. In **Early Preemption**, the backup process (the one via which Bob would have brought about the deer's death if Alice hadn't shot) is cut short early in proceedings. Once Alice shoots, there's no possibility of Bob's shooting (because there are no bullets left in their one rifle). Since this backup process is cut short early, the effect (the deer's death) depends upon the later stages of the process via which Alice's shot brings about the effect: for instance, upon m (cf. Lewis 1986b, 200). This allows us to complete a chain of causal dependence

In contrast, in **Late Preemption** the backup process isn't cut short until the effect itself (the death) occurs. That's because it's only the effect itself that prevents Bob from shooting. This means that the effect (the death) doesn't counterfactually depend upon any part of the process via which Alice's shot brought it about (cf. Lewis 1986b, 203–4). For instance, the death doesn't counterfactually depend upon the presence of a bullet at the midpoint between Alice

---

[31] In fact, Lewis (1979, 475) claims that his semantics entails the falsity of backtracking counterfactuals only where certain physical asymmetries that ordinarily correspond to the direction of time obtain. He thinks that if these physical asymmetries broke down – as they might on a closed time-like curve – then backwards causation might occur and that his semantics would entail the truth of the corresponding backtracking counterfactuals.

and the deer since, if the bullet hadn't reached that point, it wouldn't have hit the deer and so (seeing that the deer was initially unscathed) Bob would have shot and killed it.

The counterfactual reasoning from the absence of the bullet at the mid-point to Bob's shooting and the deer's death isn't in this case the sort of 'backtracking' reasoning that Lewis takes to be illicit. That's because, in **Late Preemption**, the bullet's absence would mean that it wouldn't (later) hit the deer and so Bob (later) would have shot. By contrast, in **Early Preemption**, since Bob was prevented from shooting by Alice's very act of shooting, we need to reason backwards from the bullet's absence to Alice's (earlier) not shooting to reach the conclusion that Bob would have shot and killed the deer.

So invoking transitivity doesn't solve the problem posed by late preemption. Lewis's 1973 account of causation therefore doesn't handle late preemption correctly: it fails to diagnose the preempting cause as a cause. Can a counterfactual account of causation be given that gets cases of late preemption right? Before seeking to answer that question, let's briefly consider another sort of case that arguably poses problems for Lewis's 1973 analysis.

4.4 Symmetric Overdetermination

Symmetric overdetermination occurs where there are two or more events that (in the circumstances in which they occur) are 'enough' to bring about some effect so that, in the absence of either one, the effect would still have occurred. This means that the effect doesn't counterfactually depend upon either. Unlike preemption cases – which are also cases where the effect is in a sense 'overdetermined' – symmetric overdetermination cases are those where there's no distinguishing the overdeterminers into 'preempting cause' and 'preempted backup': both have equally good claim to the title 'cause'. The following is an example.

Symmetric Overdetermination

Alice and Bob are hunting deer. Each has a loaded gun. They spot a deer. They both shoot, with their bullets piercing the deer's heart simultaneously. Each bullet alone would have been sufficient to bring about the deer's death.

In **Symmetric Overdetermination**, the deer's death doesn't counterfactually depend upon Alice's shot since, in the absence of Alice's shot, Bob's shot would still have been enough to kill the deer. Nor is there a chain of counterfactual dependence connecting Alice's shot to the deer's death. In particular, no part of the process connecting Alice's shot to the deer's death (e.g. the presence of a speeding bullet at the mid-point between Alice and the deer) is such that the deer's death depends on it: after all, the process initiated by Bob's shot was complete and would have sufficed to bring about the deer's death even had some part of the Alice-process been missing. Exactly analogous reasoning shows that the deer's death neither counterfactually depends upon, nor is connected by a chain of counterfactual dependence to, Bob's shot.

Lewis (1986b, 194–5, 207, 211–12) claims that the intuition that symmetric overdeterminers are causes isn't strong. He therefore thinks it's not a significant strike against his counterfactual analysis of causation that it doesn't deliver the result that they are. But not everyone agrees with this assessment (cf. Hitchcock 2007, 522) and there are cases in which courts have recognised symmetric overdeterminers as causes.[32] Nevertheless, since late preemption cases are the most uncontroversial counterexamples to Lewis's 1973 account of causation, we'll focus on what might be done to deal with these cases, though we'll see that some of the potential solutions may help with symmetric overdetermination too insofar as this is seen as problematic for the counterfactual approach.[33]

4.5 The Fragility Response

A natural response to the problem posed by late preemption for counterfactual theories of causation (a similar response could be adopted for dealing with early preemption cases) would be to say that the event caused by the preempting cause isn't the same as the event that would have been caused by the preempted alternative. In **Late Preemption**, for instance, one might say that the death that the deer actually died is different from the death that would have occurred had Bob shot and killed the deer. For one thing, because Bob gives Alice the first opportunity to shoot and kill the deer, a death of the deer at Bob's hands would have been later than the death that actually occurred. So perhaps the actual effect does indeed counterfactually depend upon the preempting cause after all. This response might also work in at least some symmetric overdetermination cases. For instance, in **Symmetric Overdetermination**, perhaps the actual death of the deer, due to the two bullets simultaneously piercing its heart,

---

[32] E.g. *Kingston v Chicago & N. W. Railway* [191 Wis. 610, 211 N.W. 913 (1927)]. Note that regularity theories have better prospects for counting overdeterminers as causes since each overdeterminer is (in the circumstances) sufficient for the effect.

[33] There are varieties of 'overdetermination' that differ in structure from the ones we've discussed in the last three subsections (e.g. Moore 2009, 418; Kroedel 2008, 130–1) but which we lack space to discuss here. One particularly well-known variety of overdetermination not discussed in the main text is so-called 'trumping preemption' (Schaffer 2004). However, several philosophers (e.g. McDermott 2002, 89–92, 97; Halpern and Pearl 2005, 873–5; Hitchcock 2007, 512n) take the view that 'trumping preemption' is really just an instance of symmetric overdetermination, and so has limited independent diagnostic value.

is somewhat different in manner from the death that would have occurred had only one bullet been fired.

The response under consideration invokes what Lewis (1986b, 204) calls the modal 'fragility' of the effect: the notion that a counterfactual version of the effect (e.g. the death of the deer at Bob's hands) isn't to be identified with the actual effect if it is even slightly different in time or manner from the actual effect. Lewis doesn't avail himself of this solution because he rejects extreme standards of event fragility (the adopted standards would have to be fairly extreme if this solution were to work for all cases of late preemption, including ones in which the preempted alternative would have brought about a very similar and only very slightly later version of the effect). The key reason for his doing so is that adopting extreme standards of fragility appears to yield spurious cases of causation on the counterfactual approach. For example, such things as the precise angle at which the deer was standing or even its precise blood pressure might have made a difference to the precise time and/or manner of its death, even if they were entirely irrelevant to whether or not it died. Yet it's tempting to think that such factors are not causes.[34]

4.6 De Facto Dependence and Causal Modelling

Perhaps the most influential recent attempt to modify the counterfactual approach to causation to deal with some of the challenges described here appeals to a notion that (Yablo 2002, 2004) calls *de facto* dependence. This is a sort of latent counterfactual dependence that's revealed by holding fixed certain features of the actual world.

Consider **Early Preemption**. In this example, the deer's death doesn't straightforwardly counterfactually depend upon Alice's shot. But rather than appeal to the existence of a chain of counterfactual dependence *a la* Lewis (and hence – as we've seen – commit to the rather dubious thesis that causation is transitive), the de facto dependence account says that the status of Alice's shot as a cause is revealed by the truth of the following counterfactual: *if Alice hadn't shot, and Bob still hadn't shot, then the deer wouldn't have died.* Bob's non-shooting is a feature of the actual world (or rather the world in which the

---

[34] Lewis (2004b) later came to advocate a counterfactual analysis of causation that appeals to event fragility in a subtler way. His guiding idea is that a cause 'influences' (i.e. makes a difference to) the timing and manner of its effect in a way that non-causes – such as preempted alternatives – don't. Though we lack the space to discuss the details here, this later account suffers from serious counterexamples, as Bigaj (2012) shows, and has not widely caught on.

scenario described in **Early Preemption** plays out) and the de facto depend-ence approach tells us to hold this feature fixed in seeking latent counterfactual dependence of the deer's death on Alice's shot.[35]

The counterfactual that reveals this de facto dependence of the deer's death on Alice's shot has a more complex antecedent than those to which Lewis appeals in his analysis of causation. As well as counterfactually varying the putative cause (Alice's shooting), the antecedent serves to hold fixed a certain feature of the actual situation (Bob's non-shooting). This feature was something that depended upon the putative cause (because Bob intended to shoot if Alice didn't) and therefore would have been absent under the simple counterfactual supposition that the cause didn't occur.

Hitchcock (2001a, 275) articulates this point well when he points out that Lewis – in virtue of supposing that the counterfactuals relevant to analysing causation don't *backtrack* – takes history prior to the occurrence of the event c to be implicitly held fixed (in virtue of the semantics for the counterfac-tual) when evaluating $\neg O(c) \;\Box\!\!\rightarrow\; \neg O(e)$. But Lewis (1979) takes the relevant semantics to allow *foretracking*, so that events lying to the future of c are allowed to vary under the counterfactual supposition that $\neg O(c)$. Yet, as Hitch-cock (2001a, 275) observes,' 'The necessity of foretracking from c to e does not prevent us from entertaining a counterfactual of the form 'If c had not occurred, but e had occurred anyway, then …'. Let us call a counterfactual of this sort an *explicitly nonforetracking* or ENF counterfactual.' In order to give a full-fledged account of causation in terms of ENF counterfactuals/ de facto dependence, we need an account of *which features* of the actual world should be held fixed. Several authors (e.g. Hitchcock 2001a, 2007; Halpern and Pearl 2005; Halpern and Hitchcock 2015; Halpern 2016; Weslake 2020) have appealed to *causal models* (more precisely: *structural equation models*) in giving such an account.

4.6.1 Structural Equation Models

A *structural equation model* (*SEM*), $\mathcal{M}$ comprises a set $\mathcal{V}$ of variables and a set $\mathcal{E}$ of structural equations. The values of the variables (or sometimes – as we'll see in Section 4.6.5 – the variables themselves) are taken to be, or at least to represent, potential causal relata. For example, if one wanted to model **Early Preemption**, one might include in one's model a binary variable A that takes the value $A = 1$ if Alice shoots and $A = 0$ otherwise; a binary variable *BI* that

---

[35] Though we won't go into the detals, Hitchcock (2001a) demonstrates that such an approach doesn't commit one to causal transitivity.

takes $BI = 1$ if Bob intends to shoot iff Alice doesn't and $BI = 0$ otherwise;[36] a binary variable B that takes $B = 1$ if Bob shoots and $B = 0$ otherwise; and a binary variable D that takes $D = 1$ if the deer dies, and $D = 0$ if the deer survives.

The structural equations in a model represent the way in which the model's variables depend upon one another. Specifically, each of the variables in the variable set $\mathcal{V}$ appears on the left-hand side (LHS) of exactly one structural equation in the equation set $\mathcal{E}$: the structural equation 'for' the variable in question. The variables in $\mathcal{V}$ comprise two mutually exclusive subsets: a set $\mathcal{U}$ of *exogenous variables*, the values of which don't depend upon the values of any other variables in the model, and a set $\mathcal{Y}$ of *endogenous variables*, the values of which *do* depend upon the values of other variables in the model. For example, taking the variable set of our model representing **Early Preemption** to be $\{A, BI, B, D\}$ where the variables have the interpretation given above, D is an endogenous variable, since what happens to the deer depends upon what Alice and Bob do. Variable B is also an endogenous variable, since whether Bob shoots depends upon whether Alice shoots and also upon Bob's intention. Variable A, on the other hand, is an exogenous variable since what Alice does doesn't depend either upon what Bob intends to do or what he in fact does.[37] Likewise BI is an exogenous variable because Bob's intention to shoot if Alice does not doesn't depend upon what Alice does.

The structural equation for each endogenous variable in a SEM shows how its value depends upon the values of other variables in the variable set. More precisely, where X and Z are variables in the variable set $\mathcal{V}$ of a model, X appears on the right-hand side (RHS) of the equation for Z just in case there's some assignment of values to the other variables in $\mathcal{V}$ such that the value of Z depends upon that of X if we assume those other variables are held fixed at the assigned values. Less formally, a structural equation expresses *how* the variable on the LHS depends on those variables in the model upon which it *immediately depends*: if Z immediately depends upon X and also depends on Y, but only depends on Y because Y has a bearing on the value of X, then the structural equation for Z expresses its value as a function of X but not Y. In the case

---

[36] For simplicity, I'll assume that the alternative to Bob's having this intention is his not intending to shoot under any circumstances. However, if we wanted to model the possibility that (e.g.) Bob has the intention to shoot come what may, we could deploy a multi-valued variable.

[37] Of course, Alice's shooting might depend upon Bob's not having shot *even earlier*, but if we wanted to represent whether Bob shot even earlier, we'd need an additional variable in our model. (Variable B can't serve this function, since B represents a state that depends on A. Given that this is a token causal scenario, A itself doesn't depend upon the state represented by B. If it did, we'd have a causal loop.) In general, whether a variable is exogenous or endogenous depends on what other variables we include in our model.

of an exogenous variable, W, no variable appears on the RHS of the equation for W. Rather, the equation simply takes the form $W = w^*$, where w^* is the actual value of W. Any variables that appear on the RHS of the equation for a variable V are known as the 'parents' of V; V is a 'child' of theirs. The notions of 'ancestor' and 'descendant' are defined in terms of the transitive closures of parenthood and childhood respectively.

Thus, in our model of **Early Preemption**, the structural equations for A and BI just state their actual values, $A = 1$ and $BI = 1$, because A and BI are exogenous. Since the value of B (immediately) depends upon the value of A and the value of BI (and not the value of D: no backtracking!), the equation for B expresses its value just as a function of A and BI. Specifically, the equation is $B = Min\{BI, 1 - A\}$ which encodes the fact that Bob shoots ($B = 1$) iff he has the intention ($BI = 1$) and Alice doesn't shoot ($A = 0$). Finally, the structural equation for D expresses its value as a function of A and B, but not of BI. That's because D immediately depends upon A and B, but doesn't immediately depend upon BI. This is evidenced by the fact that there's no combination of values of A and B such that the value of D depends upon that of BI when A and B are held fixed at those values: specifically, once we know whether Bob shoots, what he earlier intended has no further bearing on the deer's death. (By contrast, it's easy to verify that there are values of A and BI – e.g. $A = 0$ and $BI = 1$ – such that the value of D depends upon the value of B when A and BI are held fixed at those values, and that there are values of B and BI – e.g. $B = 0$ and $BI = 1$ – such that the value of D depends upon the value of A when B and BI are held fixed at those values.) The structural equation for D is, specifically, $D = Max\{A, B\}$, reflecting the fact that D takes value $D = 1$ (i.e. the deer dies) iff either $A = 1$ (Alice shoots) or $B = 1$ (Bob shoots). To summarise, our model '$\mathcal{EP}$' of **Early Preemption** is as follows:

Variables: $\{A, BI, B, D\}$

Equations: $A = 1$; $BI = 1$; $B = Min\{BI, 1 - A\}$; $D = Max\{A, B\}$

The solution to these structural equations is $A = 1$, $BI = 1$, $B = 0$, and $D = 1$. That is, Alice shoots, Bob intends to shoot iff Alice doesn't, Bob doesn't shoot, and the deer dies.

Since structural equations encode information about counterfactual dependence (specifically, about how the variable on the LHS counterfactually depends upon the variables on the RHS), they differ from algebraic equations. That's because counterfactual dependence isn't symmetric – it doesn't 'backtrack' (at least not on the sort of counterfactual semantics that we need if we're to successfully analyse causation in terms of counterfactuals) – and so algebraic rearrangement of structural equations won't generally preserve truth.

A SEM can be given a graphical representation by taking the variables in $\mathcal{V}$ as the nodes of the graph and drawing a directed edge ('arrow') from a variable V_i to a variable V_j ($V_i, V_j \in \mathcal{V}$) just in case V_i is a parent of V_j according to the structural equations in $\mathcal{E}$. Implementing this rule for $\mathcal{EP}$ yields the graph depicted in Figure 2 (which follows a summary of the model $\mathcal{EP}$). A 'directed path' can then be defined as an ordered sequence of variables $\langle V_i, V_j, \ldots, V_k \rangle$, such that there's a directed edge from V_i to V_j, and a directed edge from V_j to $\ldots V_k$ (i.e. directed paths run from variables to their descendants). The directed paths in the graph depicted in Figure 2 are thus $\langle A, D \rangle$, $\langle BI, B \rangle$, $\langle A, B \rangle$, $\langle B, D \rangle$, $\langle BI, B, D \rangle$, and $\langle A, B, D \rangle$.

$\mathcal{EP}$

Variables: $\{A, BI, B, D\}$

Interpretation: $A = 1/0$ if Alice shoots/doesn't shoot; $BI = 1/0$ if Bob intends/doesn't intend to shoot iff Alice doesn't shoot; $B = 1/0$ if Bob shoots/doesn't shoot; $D = 1/0$ if the deer dies/survives.

Equations: $A = 1$; $BI = 1$; $B = Min\{BI, 1 - A\}$; $D = Max\{A, B\}$

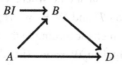

Figure 2: Early Preemption

One can evaluate a counterfactual of the form $V_i = v_i \& \ldots \& V_k = v_k \;\square\!\!\rightarrow Z = z$ with respect to a SEM by replacing the equations for $V_i, \ldots,$ and V_k with the equations $V_i = v_i, \ldots,$ and $V_k = v_k$ (thus rendering each of $V_i, \ldots,$ and V_k exogenous), while leaving all other equations intact. The result is a new set of equations. The counterfactual holds just in case, in the solution to the new set of equations, $Z = z$. This 'equation replacement' method gives us a means to evaluate even those counterfactuals whose truth or falsity isn't implied by any single equation in the model alone (Hitchcock 2001a, 283): for example, counterfactuals saying how the value of a variable would differ if the values of its grandparents were different.[38]

---

[38] As we'll discuss further in Section 4.6.6, the equation replacement method for evaluating counterfactuals is sometimes taken as modelling what would happen if the variables in the antecedent

So, for example, if one wants to evaluate the ENF counterfactual $A = 0\&B = 0 \,\square\!\!\rightarrow D = 0$ – which is the one that advocates of the de facto dependence approach take to be indicative of the causation of $D = 1$ by $A = 1$ in **Early Preemption** – with respect to our model $\mathcal{EP}$, then one needs to replace the equation that $\mathcal{EP}$ gives for A (i.e. $A = 1$) with the equation $A = 0$ and the equation that $\mathcal{EP}$ gives for B (i.e. $B = Min\{BI, 1 - A\}$) with the equation $B = 0$ while leaving the equations that $\mathcal{EP}$ gives for BI (i.e. $BI = 1$) and D (i.e. $D = Max\{A, B\}$) intact. In the solution to this new set of equations, $D = 0$, hence the counterfactual is true relative to $\mathcal{EP}$.

4.6.2 A Structural Equations Analysis of Causation

A simple and elegant proposal for analysing causation in terms of SEMs is given by **SEA** (for Structural Equations Analysis). This analysis is closely related – but not identical – to the analysis given by Hitchcock (2001a, 286–7).

SEA

Where x and x' ($x \neq x'$) are possible values of the variable X and y and y' ($y \neq y'$) are possible values of the variable Y, then X taking value $X = x$ rather than $X = x'$ is a cause of Y taking value $Y = y$ (rather than $Y = y'$) iff $X = x$ and $Y = y$ are the actual values of X and Y and, in some apt SEM, there's a path $\mathcal{P}$ from X to Y such that, where **W** is the set of variables in the model that don't lie on $\mathcal{P}$ and $\mathbf{W} = \mathbf{w}^*$ represents the variables **W** taking their actual values, then (according to the model) if X had taken the value $X = x'$ and **W** had taken $\mathbf{W} = \mathbf{w}^*$ (i.e. the off-path variables had taken their actual values), then Y would have taken the value $Y = y'$.

There's quite a bit to unpack in this analysis. First, it analyses a contrastive notion of causation. As noted in Section 3.2, taking causation to be contrastive seems necessary when dealing with multi-valued variables. When dealing with binary variables, the contrast is obvious and so mention of it can be suppressed (and often will be in what follows). Second, the reference to 'apt' models. We've seen what a SEM is, but what makes such a model apt? We'll return to this issue in the next subsection, but let's assume for now that the model $\mathcal{EP}$ *is* apt (spoiler: it passes all tests for appropriateness that have been suggested in the literature) and see how **SEA** gives the verdict that $A = 1$ (rather than

---

of the counterfactual were set to the specified values by means of 'interventions' (Woodward 2003, 98). Glynn (2013) discusses the extent to which the equation replacement method could alternatively be seen as modelling a version of Lewis's (1979) counterfactual semantics.

$A = 0$) causes $D = 1$ (rather than $D = 0$): that is, that Alice's shooting (rather than not) causes the deer to die (rather than survive).

To see this, focus on the path $\langle A, D \rangle$ in $\mathcal{EP}$. **SEA** tells us that $A = 1$ is a cause of $D = 1$ if it's true that, when all variables that lie off this path are held fixed at their actual values, if A had taken the value $A = 0$, then D would have taken the value $D = 0$. In $\mathcal{EP}$, there are two variables that lie off the path $\langle A, D \rangle$: namely BI and B, which have $BI = 1$ and $B = 0$ as their actual values. Thus the fact that $A = 1$ is a cause of $D = 1$ is indicated by the truth relative to $\mathcal{EP}$ of the ENF counterfactual $A = 0 \& BI = 1 \& B = 0 \mathbin{\square\!\!\rightarrow} D = 0$.[39] The truth of this ENF counterfactual relative to $\mathcal{EP}$ can be verified by employing the equation replacement method described in the previous subsection.

Analysis **SEA** also yields the correct verdict that $BI = 1$ *isn't* a cause of $D = 1$ (confirming that it's 'merely' a preempted backup). To see this, consider the sole directed path from BI to D in $\mathcal{EP}$: $\langle BI, B, D \rangle$. The sole variable that lies off this path is A, the actual value of which is $A = 1$. For $BI = 1$ to count as a cause of $D = 1$ according to **SEA**, it would be necessary that the following counterfactual hold in $\mathcal{EP}$: $BI = 0 \& A = 1 \mathbin{\square\!\!\rightarrow} D = 0$. However it's easy to verify via the equation replacement method that this isn't the case.

Analysis **SEA** can also potentially help with **Late Preemption**,[40] an example which – as we saw – Lewis struggles to deal with. To model **Late Preemption** it will be helpful to recall that, since Bob first waits to see whether Alice succeeds in killing the deer, the deer would have died slightly later if Alice hadn't killed it. Suppose the deer would have died by 1:00pm if Alice had shot, but only by 1:01pm if Alice hadn't shot. Thus we can adopt the following model ('$\mathcal{LP}$') of **Late Preemption**.[41]

---

[39] I previously said that $A = 0 \& B = 0 \mathbin{\square\!\!\rightarrow} D = 0$ is the counterfactual that the advocate of the de facto dependence approach takes to be indicative of causation of $D = 1$ by $A = 1$. And so it would be, according to **SEA**, if our model only included variables A, B, and D. But what I said previously was a slight simplification in that, once BI is included in our model, it too must be held fixed according to **SEA**'s general recipe for revealing the sort of de facto dependence that's indicative of causation. In this instance, holding fixed BI – although harmless – isn't doing any real work. It's the holding fixed of B at $B = 0$ that's the key to revealing latent dependence of $D = 1$ on $A = 1$.

[40] The treatment suggested here follows that of Halpern and Pearl (2005, 863–5) and Hitchcock (2007, 524–9).

[41] It's worth emphasising how the interpretation of variable B differs in $\mathcal{LP}$ as compared to $\mathcal{EP}$. In $\mathcal{LP}$, $B = 1$ represents Bob's intention to shoot *iff Alice doesn't kill the deer* (rather than iff Alice doesn't shoot). As in the case of $\mathcal{EP}$, I assume for simplicity that the alternative to Bob's having the intention he does is for him to have no intention to shoot the deer under any circumstances.

$\mathcal{LP}$

Variables: $\{A, BI, B, D_{1:00}, D_{1:01}\}$

Interpretation: $A = 1/0$ if Alice shoots/doesn't shoot; $BI = 1/0$ if Bob intends/doesn't intend to shoot iff Alice *doesn't kill the deer*; $B = 1/0$ if Bob shoots/doesn't shoot; $D_{1:00} = 1/0$ if the deer is/isn't dead by 1:00pm; $D_{1:01} = 1/0$ if the deer is/isn't dead by 1:01pm.

Equations: $A = 1$; $BI = 1$; $D_{1:00} = A$; $B = Min\{BI, 1 - D_{1:00}\}$; $D_{1:01} = Max\{D_{1:00}, B\}$

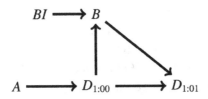

Figure 3: Late Preemption

Assuming $\mathcal{LP}$ is apt then, according to **SEA**, $A = 1$ is a cause both of $D_{1:00} = 1$ and $D_{1:01} = 1$. First consider the sole path between A and $D_{1:00}$: $\langle A, D_{1:00} \rangle$. There are three variables that lie off this path: BI, B, and $D_{1:01}$ with the actual values $BI = 1$, $B = 0$, and $D_{1:01} = 1$. **SEA** counts $A = 1$ as a cause of $D_{1:00} = 1$ in virtue of the truth of the following counterfactual relative to $\mathcal{LP}$: $A = 0 \& BI = 1 \& B = 0 \& D_{1:01} = 1 \ \Box\!\!\rightarrow\ D_{1:00} = 0$. The truth of this counterfactual is easily verified using the equation replacement method. This exhibits the fact that the equation replacement method corresponds to a non-backtracking way of evaluating counterfactuals. In particular, it doesn't validate the reasoning that if the deer had been dead at 1:01pm, then it must have been dead at 1:00pm. Indeed it doesn't validate this reasoning even on the assumption that neither Alice nor Bob shot.

When it comes to the relationship between $A = 1$ and $D_{1:01}$, consider the path $\langle A, D_{1:00}, D_{1:01} \rangle$. There are two variables – BI and B – that lie off this path and their actual values are $BI = 1$ and $B = 0$. **SEA** implies that $A = 1$ is a cause of $D_{1:01} = 1$ if the latter depends upon the former when we hold fixed BI and B at their actual values: that is, if it's true that $A = 0 \& BI = 1 \& B = 0 \ \Box\!\!\rightarrow\ D_{1:01} = 0$. That this is true can again be verified via the equation replacement method. Although the event of the deer's death isn't straightforwardly represented in $\mathcal{LP}$ – though one possibility would be to take it to correspond to the disjunction $D_{1:00} = 1 \lor D_{1:01} = 1$ (cf. Hitchcock 2007, 525) – following Hitchcock (2007,

525), we might suppose that 'we will have adequately captured our intuitions about the case if we can show that' $A = 1$ is a cause of both $D_{1:00} = 1$ and $D_{1:01} = 1$.

It's also easy to see that $BI = 1$ isn't a cause of $D_{1:00} = 1$ or $D_{1:01} = 1$ according to **SEA**.[42] First, since there's not even a directed path from BI to $D_{1:00}$ (reflecting the fact that Bob is going to give Alice until at least 1:00pm to kill the deer before himself shooting), **SEA** straightforwardly doesn't count $BI = 1$ as a cause of $D_{1:00} = 1$. Second, though there's a directed path from BI to $D_{1:01}$ – namely, $\langle BI, B, D_{1:01} \rangle$ – holding fixed the off-path variables A and $D_{1:00}$ at their actual values – $A = 1$ and $D_{1:00} = 1$ – doesn't reveal de facto dependence of $D_{1:01} = 1$ upon $A = 1$: it's false that $BI = 0 \& A = 1 \& D_{1:00} = 1 \ \Box\!\!\rightarrow D = 0$, as can again be verified via the equation replacement method.

Although **SEA** deals well with **Early Preemption** and at least reasonably well with **Late Preemption**, it arguably runs into difficulties with **Symmetric Overdetermination**. Before demonstrating this and discussing whether there might be an alternative SEM analysis of causation that fares better, we ought to return to the question – deferred earlier – of what makes for an apt SEM.

4.6.3 Apt Models

Several criteria for model appropriateness have been suggested in the literature. Specifically, the following criteria for the variables of such a model have been suggested:

1. (Partition) The values of each of the variables in the variable set ought to form a partition (i.e. a set of mutually exclusive and jointly exhaustive alternatives) (Halpern and Hitchcock 2010, 397–8; Blanchard and Schaffer 2015, 182).
2. (Independence) The values of distinct variables in the variable set shouldn't be logically or metaphysically related (Hitchcock 2001a, 287; Halpern and Hitchcock 2010, 397).
3. (Naturalness) The values of the variables ought to represent reasonably natural and intrinsic states of affairs (cf. Blanchard and Schaffer 2015, 182).

Since analyses like **SEA** take all and only values of distinct variables to be potential causal relata, (Partition) ensures that we don't thereby miss actual causal relations because they obtain between the values of a single variable,

---

[42] At least **SEA** doesn't regard $BI = 1$ as a cause of $D_{1:00} = 1$ or $D_{1:01} = 1$ *relative to the model $\mathcal{LP}$*. Yet **SEA** existentially quantifies over apt models so $BI = 1$ will count as a cause *simpliciter* of $D_{1:00} = 1$ or $D_{1:01} = 1$ if it's a cause *relative to at least one apt model*. In the next subsection, we'll discuss the extent to which we have reason to think that models relative to which $BI = 1$ counts as a cause of $D_{1:00} = 1$ or $D_{1:01} = 1$ are *inapt*.

(Independence) ensures that we don't mistake stronger-than-causal relations for causal relations, and (Naturalness) ensures that unnatural or non-intrinsic states of affairs do not get counted as causes and effects (justification for such a restriction on causal relata is given by Lewis 1986b, 190, 263).[43] Three other conditions of model appropriateness are also fairly standard:

4. (Veridicality) The model should entail only true counterfactuals (Hitchcock 2001a, 287; 2007, 503)
5. (Serious Possibilities) The model shouldn't represent possibilities "that we consider to be too remote" (Hitchcock 2001a, 287; cf. Woodward 2003, 86–91; Weslake 2020)
6. (Stability) It shouldn't be possible to overturn causal verdicts reached relative to the model by moving to a model (satisfying Partition, Independence, Naturalness, Veridicality, and Serious Possibilities) that's richer in the sense of having a variable set that's a superset of that of the original model (or perhaps by including variables that are fine-grainings of those in the original model) (cf. Halpern and Hitchcock 2010, 394–5; Halpern 2015; Hitchcock 2007, 503).

The motivation for (Veridicality) should be obvious: if the model entails false counterfactuals (via the equation replacement method described in Section 4.6.1), then it's not a reliable guide to genuine causal relations. (Serious Possibilities) is needed to avoid a variety of counterintuitive results (see, e.g., Hitchcock 2001a, 287; Woodward 2003, 86–91). Most straightforwardly, it helps avoid the overgeneration of cases of causation by absence. For instance, JFK's death presumably counterfactually depends upon Paul McCartney's failure to get in the way of the oncoming bullets. But most of us would be reluctant to say that the latter is a cause of the former. This might be explained by the fact that we view McCartney's getting in the way as a non-serious possibility. (Serious Possibilities) ensures that no appropriate model of the events surrounding

---

[43] The fact that (Naturalness) refers to 'reasonably' natural and intrinsic states of affairs introduces a certain amount of vagueness into the account, but is necessary because demanding perfect naturalness would presumably rule out high-level states of affairs (i.e. the states – such as price inflation and asbestos exposure – of concern to the non-fundamental sciences) as causes and effects, which would be undesirable. Since *any* account of causation will need a condition like this, the SEM approach doesn't suffer a relative disadvantage in this regard.

If *absences* are unnatural states of affairs (e.g. if they're disjunctions of positive events, cf. Lewis 1986b, 189–93), we might replace (Naturalness) with the requirement that each variable has *at most one value* representing such a state of affairs if we wish to allow (as we will in the following) that absences of reasonably natural events should count as causes and effects, and also that positive events that are causes and effects may be represented by binary variables that have the absence of these events as one of their values.

JFK's death includes a variable representing whether McCartney gets in the way. Consequently no such model will treat his failure to do so as a cause.

Adopting (Serious Possibilities) results in a certain amount of vagueness and, perhaps, subjectivity. But traditional accounts of causation – which don't appeal to causal models – also stand in need of appeal to serious possibilities because, for instance, McCartney's failure to get in the way is an inus condition of JFK's death and the latter counterfactually depends upon the former (cf. Collins 2000, esp. 229; Woodward 2003, 86–8; McGrath 2005).[44]

The idea behind (Stability) is that an appropriate model is a sufficiently rich representation of causal reality that moving to a richer representation by including additional variables representing further factors (and/or by representing already-represented factors in more detail) wouldn't reveal causal verdicts to be spurious.[45] This requirement just reflects the obvious fact that we don't want to make causal judgements on the basis of a model that's too impoverished to accurately reflect causal reality.

It should be easy for the reader to verify that the models $\mathcal{EP}$ and $\mathcal{LP}$, as well as the models to be considered in the remainder of this section, satisfy (Partition), (Independence), and (Veridicality) (where the latter is assessed relative to the worlds of the scenarios that they model) and that they don't represent overly gerrymandered or extrinsic states, thus satisfying (Naturalness). As already noted (Serious Possibilities) is vague, but most participants in the literature accept standards for what counts as a 'serious possibility' that don't rule either $\mathcal{EP}$ or $\mathcal{LP}$ as inapt. Specifying exactly what these standards amount to is tricky: after all, there are combinations of possible values for the variables that figure in $\mathcal{EP}$ and $\mathcal{LP}$ that correspond to intuitively weird possibilities – e.g. that Alice and Bob shoot but the deer doesn't die or, weirder still, in the case of $\mathcal{LP}$, that the deer is dead at 1:00pm but not at 1:01pm. But let's assume for the sake of argument there's a way of cashing out (Serious Possibilities) so that (as advocates of the SEM approach seem to desire) models like $\mathcal{EP}$ and $\mathcal{LP}$ don't count as violating it, but (say) a model of JFK's assassination that includes a variable corresponding to McCartney's jumping in the way of the bullets does.[46]

---

[44] If one regards appeal to 'serious possibilities' in an account of causation as too problematic, one option would be to simply admit that, e.g., McCartney's failure to get in the way *was* a cause of JFK's death. If so, as Blanchard and Schaffer (2015, 198) point out, (Serious Possibilities) 'may be reinterpreted, not as an aptness constraint on models, but as a descriptive psychological claim about which causal models are most readily available to us when we form our causal judgements'.

[45] (Stability) thus renders the notion of an appropriate model relative to the causal claim being evaluated.

[46] For further discussion, see Hitchcock (2001a, 295–8).

Convincing ourselves that (Stability) is satisfied is also something of a difficult task: it doesn't seem that there can be any *proof* of the non-existence of unmodelled variables whose inclusion would overturn causal verdicts. This could be seen as a drawback of the SEM approach. On the other hand, one might argue that the discovery of relevant but previously unmodelled variables is precisely the sort of thing that *should* lead us to reconsider our causal verdicts and that, as any scientist will tell you, it's just a fact of life that we can never attain total certainty that all critical variables have been taken into account. In any case, given the information we're provided with in **Early Preemption** and **Late Preemption**, it doesn't seem that there's any way of enriching $\mathcal{EP}$ or $\mathcal{LP}$ to undermine the causal verdicts that we reached. In what follows, I'll occasionally comment on cases where there seem to be obvious enrichments of the models that we consider and evaluate whether causal verdicts are overturned when these enrichments are made.

Of course one thing that ought to trouble the advocate of the structural equations approach is if there are systematic biases in the models that occur to us. This could potentially render our judgements about the likely stability of a given model unreliable. There are at least two potential sources of bias that have been pointed out in the literature. First, Glymour et al. (2010) have pointed to an excessive focus on simple models with small numbers of variables. Such a focus likely reflects the cognitive limitations of humans. Second, it might be that our *causal presuppositions* influence our model construction (Paul and Hall 2013, 4.2.8–9).[47] Whilst we don't have space to do justice to this topic here, it seems a rich vein for further research that is liable to be interdisciplinary both in its nature (since the input of specialists in psychology and machine learning would appear to be crucial) and in its implications (since it's relevant to an assessment of the objectivity of model-building in the sciences).

4.6.4 A Problem for SEA

Even if **SEA** fares well with respect to **Early Preemption** and **Late Preemption**, it runs into difficulties with **Symmetric Overdetermination**. The trouble is that, in this case, there doesn't appear to be any feature of the actual situation that we can hold fixed to reveal de facto dependence. The following model ('$\mathcal{SO}$') would seem to be a reasonable one for **Symmetric Overdetermination**:

---

[47] Paul and Hall focus in particular upon the influence that such presuppositions have on our judgements about relations between variables (i.e. what structural equations we judge to be true). This is liable to affect our judgements about (Veridicality) as well as (Stability).

$\mathcal{SO}$

Variables: $\{A, B, D\}$

Interpretation: $A = 1/0$ if Alice shoots/doesn't shoot; $B = 1/0$ if Bob shoots/doesn't shoot; $D = 1/0$ if the deer dies/survives.

Equations: $A = 1$, $B = 1$, $D = Max\{A, B\}$

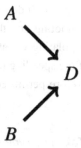

Figure 4: Symmetric Overdetermination

There's only one path from A to D and only one variable off this path, namely B. So the test for causation appealed to by **SEA** requires that $D = 1$ de facto depends upon $A = 1$ when B is held fixed at its actual value $B = 1$. But there's no such dependence: the deer's death doesn't counterfactually depend upon Alice's shot when we hold fixed Bob's shot. The falsity of the counterfactual $A = 0 \& B = 1 \:\Box\!\!\rightarrow\: D = 0$ can be verified via the equation replacement method (described in Section 4.6.1) with respect to $\mathcal{SO}$. Strictly analogous considerations show that **SEA** yields the verdict that $B = 1$ isn't a cause of $D = 1$ either. This is a problem to the extent that we regard symmetric overdeterminers as causes.

The most obvious way to try to deal with **Symmetric Overdetermination** is to liberalise the analysis by allowing that we may sometimes *vary* features of the actual situation in seeking latent dependence of effect on putative cause. This is an approach taken by Hitchcock (2001a, 289) and Halpern and Pearl (2005), inter alia. For instance, were we allowed to suppose that Bob *didn't* shoot (and hold this fixed by including it in the antecedent of the relevant ENF), then we'd recover counterfactual dependence of the deer's death on Alice's shooting.

But we don't want to be too liberal: if we're allowed to vary just any feature of the actual situation, we'll end up generating spurious cases of causation. For instance, consider the intention of Bob to shoot if Alice doesn't kill the deer in **Late Preemption**. If we hold fixed the non-actual fact that Alice didn't shoot (by including it in the antecedent of the relevant ENF), then the deer's death comes to counterfactually depend upon Bob's intention. Yet Bob's intention isn't a cause of the deer's death.

Briefly, the way this problem is standardly tackled (Hitchcock 2001a, 289; Halpern and Pearl 2005, 853–5) is by noting that, in **Late Preemption**, if Alice hadn't shot ($A = 0$), then the causal path between Bob's intention and the deer's death would have been crucially different: Bob would have shot rather than not doing so (i.e. B would have taken $B = 1$ rather than $B = 0$ in $\mathcal{LP}$). In contrast, in $\mathcal{SO}$, setting $A = 0$ doesn't make any difference to the causal pathway between Bob's shooting and the deer's dying (and this is so even if we enriched $\mathcal{SO}$ by interpolating variables between B and D, such as one representing the presence of the speeding bullet mid-air between Bob and the deer). Setting $A = 0$ is therefore permitted in the latter case but not the former.

4.6.5 Token vs Type Again

It was remarked at the outset of this section that counterfactual analyses of causation have typically been directed at understanding token causation. Having said that, when it comes to approaches that invoke variables, it's possible to analyse a family of causal relations that aren't in themselves type or token causal relations, but in terms of which it might be possible to analyse type causation: namely, a family of relations of causation between *variables* as opposed to the (actual) *values* of those variables. Causal relations between variables are discussed quite extensively by Pearl (2009), Spirtes et al. (2000), and Woodward (2003).

Woodward (2003) distinguishes various relations of causation between variables. *Direct causation* (Woodward 2003, 55), which is an inherently model-relative notion of causation, is defined by him so that X counts as a *direct cause* of Y in a model iff X is a parent of Y. On the other hand, X is defined as a *total cause* of Y iff there's a possible change to the value of X that makes a difference to the value of Y (Woodward 2003, 51).[48] This isn't an inherently model-relative notion of causation.

Finally, X is defined as a *contributing cause* of Y relative to a model iff there's a path $\mathcal{P}$ from X to Y and some possible setting of the off-path variables (which doesn't have to be their actual values) such that, when the off-path variables are

---

[48] Formally: X is a *total cause* of Y iff there's a pair of values x, x' ($x \neq x'$) of X (neither of which need be the actual value) and a pair of values y, y' ($y \neq y'$) of Y such that, had X taken $X = x$, then Y would have taken $Y = y$ and had X taken $X = x'$, then Y would have taken $Y = y'$.

held fixed at that setting, there's a possible change to the value of X that makes a difference to the value of Y (Woodward 2003, 57).[49] Though this definition is model-relativised, one might define X to be a contributing cause of Y *simpliciter* iff X is a *contributing cause* of Y relative to at least one apt model.[50]

These definitions of various causal relations that might obtain between variables don't invoke the actual values of the cause and effect variables. This sets them apart from definitions of actual causation (i.e. the token causal relation that has been of central philosophical focus). Still, they don't necessarily capture type-causal notions either (cf. Hitchcock 2007, 503–4). Specifically, there's no requirement that the variables in question represent type-level phenomena. For example, according to these definitions, A counts as a direct, total, and contributing cause of $D_{1:00}$ in our model $\mathcal{LP}$ of **Late Preemption**. Yet A and $D_{1:00}$ represent quite specific things: namely, whether Alice shoots when she has the chance, and whether the deer is dead at 1pm. In general, where variables represent particulars such as token events, causation between such variables isn't 'type causation' in the usual sense. Hitchcock (2007, 503–4) suggests that we might in such cases regard these relations as part of what he calls the 'token causal structure' of a scenario, but not as instances of actual causation.[51]

Still, it's of course possible to use variables to represent things that are nonparticular. For instance, consider the Arrhenius Equation (AE) in chemistry. The AE shows how the rate of a chemical reaction depends upon temperature. It can be formulated as follows:

$$k = A exp^{\frac{-E_a}{RT}} . \tag{AE}$$

---

[49] Formally: X is a *contributing cause* of Y iff there's a pair of values x, x' ($x \neq x'$) of X (neither of which need be the actual value), a pair of values y, y' ($y \neq y'$), and an assignment of values to all off-path variables such that (i) had X taken $X = x$ and had the off-path variables taken the assigned values, then Y would have taken $Y = y$ and, (ii) had X taken $X = x'$ and had the off-path variables taken the assigned values, then Y would have taken $Y = y'$.

[50] Woodward (2003, 51, 57) provides probabilistic versions of his definitions of *total* and *contributing* cause, which will be considered in Section 5.6.

[51] Of course, where the variables in our model concern particulars, we could build on Woodward's definitions to define a variety of 'token causal relations'. For instance if x and y were required to be the actual values of X and Y in the definition of 'total cause' given in Footnote 48, then we'd have a 'token causal relation' that was defined in terms of straightforward counterfactual dependence (though explicitly incorporating contrastivity). Still, as should be clear in light of the foregoing subsections, none of these definitions (even when restricted to focus on the actual values of X and Y and the requirement that the variables concern particulars) seems particularly well aligned with 'actual causation': the token causal relation that's been the subject of so much philosophical interest (as well as interest from lawyers, psychologists, and computer scientists) and that seems to be the central token causal concept of scientists and laypeople.

Here k is the reaction rate (the amount of reaction product produced per litre of the reaction solution per second), A and E_a are constants whose values depend upon the nature of the reactants, R is the universal gas constant, and T is absolute temperature. The AE captures the fact that small increases in temperature result in dramatic increases in reaction rate. AE is a single-equation SEM where k and T are the variables.

Relative to AE, temperature T counts as a direct, total, and contributing cause of the reaction rate k by Woodward's definitions. Moreover, the variables in AE don't concern particulars. They are general in the sense that the AE can be applied to any reaction rather than being limited to a single one. As applied to variables like these, Woodward's notions of causation between variables can reasonably be thought of as capturing type-level relations.

4.6.6 Interventions

Before closing our discussion of the counterfactual approach it's worth noting that, with the notion of a SEM in hand, we're now in a position to more precisely characterise the notion of an 'intervention', which plays an important role in some theorists' thinking about the SEM approach to causation, and which we'll appeal to again in our discussion of probabilistic causation. Our characterisation will be a slightly simplified version of that given by Woodward (2003, 94–114).

On Woodward's approach, interventions are characterised in terms of 'intervention variables'. The notion of an intervention variable can be understood in terms of a three-place relation between variables: 'I is an intervention variable for X with respect to Y', where X is the variable that's the subject of the intervention, and the intervention is one that can test whether Y depends upon X. We can keep track of this using subscripts: a variable that's a putative intervention variable for X with respect to Y can be denoted $I_{X,Y}$.

According to Woodward's definition, $I_{X,Y}$ counts as an intervention variable for X with respect to Y iff (a) $I_{X,Y}$ is a contributing cause of X (in the sense defined in the previous subsection); (b) $I_{X,Y}$ acts as a 'switch' for X: that is, there are values of $I_{X,Y}$ (call these 'switching values') such that when $I_{X,Y}$ takes those values, the value of X no longer depends on the values of any variables besides X that don't lie intermediate on a directed path from $I_{X,Y}$ to X; (c) there's no directed path from $I_{X,Y}$ to Y that doesn't go via X; (d) there's no variable that's an ancestor of $I_{X,Y}$ that lies on a directed path to Y that doesn't run via $I_{X,Y}$ (i.e. there's no 'common cause' of $I_{X,Y}$ and Y).[52]

---

[52] It seems there's something of a regress in the offing in the definition of an intervention variable. As we'll see, intervention variables are supposed to test what counterfactuals are true and

The idea, then, is that there are circumstances under which $I_{X,Y}$ makes a difference to the value of X (condition (a)) and indeed there's a certain set of values of $I_{X,Y}$ such that when $I_{X,Y}$ takes one among those values, the value of X doesn't depend upon those of any other variable besides $I_{X,Y}$ (except any that lie on a path from $I_{X,Y}$ to X) (condition (b)). Moreover, $I_{X,Y}$ isn't a variable that potentially influences Y independently of its influence on X: any influence that $I_{X,Y}$ has on Y goes by way of its influence upon X (condition (c)). Finally, there's no variable that's a 'common cause' of $I_{X,Y}$ and Y so there's no correlation between $I_{X,Y}$ and Y that isn't explained by the influence of $I_{X,Y}$ upon X (condition (d)). The latter two conditions are designed to ensure that any changes in the value of Y associated with changes in $I_{X,Y}$ are entirely due to changes that $I_{X,Y}$ produces in X. This ensures that any such changes in the value of Y genuinely reflect its dependence on X.

It's consistent with the definition of an intervention variable that, when the value of $I_{X,Y}$ lies outside the set of 'switching values', its value doesn't make a difference to that of X. In such circumstances, there might be apt models from which this variable can be omitted (i.e. left as latent). Of course if the variable is then 'added' to the model, this will require an adjustment to the structural equations so that the equation for X now reflects its dependence upon $I_{X,Y}$ in addition to its other parents in the model.

To illustrate, recall the example **Early Preemption**. In this scenario, the bullet's passing the mid-point between Alice and the deer might be represented by a variable M which takes $M = 1$ if the bullet indeed passes this mid-point and $M = 0$ otherwise (see Figure 5). Now imagine a putative intervention on M with respect to D (the variable representing the deer's death) that comprises someone's placing an object that would intercept the speeding bullet before it reached this mid-point. Whether such an object is placed might be represented by a binary variable $I_{M,D}$ that takes value $I_{M,D} = 1$ if one is or $I_{M,D} = 0$ otherwise.

Consider whether $I_{M,D}$ is a genuine intervention variable for M with respect to D, the variable representing whether the deer dies. $I_{M,D}$ acts as a 'switch' with respect to M in the sense that, when $I_{M,D} = 0$, the value of M depends upon the value of A – representing whether Alice shoots – but when $I_{M,D} = 1$, then $M = 0$ no matter what the value of A. Moreover, as is implied by its acting as a switch, $I_{M,D}$ is a contributing cause of M (i.e. there are circumstances under

---

therefore what the structural equations are – and hence what the arrows in the graph are – in a model satisfying (Veridicality). But don't we need an intervention variable for the intervention variable in order to establish what arrows emanate from $I_{X,Y}$ itself, and therefore whether $I_{X,Y}$ satisfies (for instance) condition (c)? This issue is taken up in (Glynn 2013).

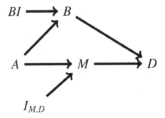

Figure 5 An Intervention Variable for M with respect to D

which it makes a difference to the value of M). Now if no one actually places an object to intercept the bullet (indeed, perhaps there's no-one present who has any intention or capacity to do so), then it would be entirely reasonable to omit $I_{M,D}$ from our model altogether. However, once $I_{M,D}$ is included (as it is in the model represented in Figure 5), it can be verified that it satisfies the remaining conditions for an intervention variable for M with respect to D. Specifically, there's no directed path from $I_{M,D}$ to D that doesn't go via M and there's no variable that's an ancestor of $I_{M,D}$ that lies on a directed path to D that doesn't run via $I_{M,D}$.[53]

Note, by contrast, that if we added a variable that was a parent of A (perhaps representing whether someone shouts at Alice to stop her from shooting) to our model then this wouldn't count as an intervention variable for M with respect to D (though – so long as such a shout wouldn't scare off the deer – it may count as an intervention variable for A with respect to D). That's because there would be a directed path from such a variable to D that bypasses M: namely the one that runs via A and B.

When it comes to conditions (c) and (d) for an intervention variable, it's of course not enough that they be satisfied in just any old model that includes $I_{X,Y}$, X, and Y in its variable set. Rather it will have to be the case that, no matter how many variables we include in our model, then (provided the resulting model satisfies (Veridicality), (Naturalness), etc.) conditions (c) and (d) still hold. That's because if there's a path from $I_{X,Y}$ to Y that bypasses X that could be revealed by the addition of more variables to the model, or if there are unmodeled common causes of $I_{X,Y}$ and Y, then changes in the value of $I_{X,Y}$ won't reliably test the influence of X upon Y.

---

[53] It might be objected that – if in fact there's no one around with the intention and capacity to place an object that would intercept the bullet – a model including $I_{M,D}$ violates (Serious Possibilities). One possible response would be to claim that, while we may wish to think about interventions (which may in some cases correspond to far-fetched possibilities) when considering whether the counterfactuals entailed by a causal model are true, and while to help us do so we may wish to expand the original model to include variables representing these interventions, it isn't the expanded model of which we need (Serious Possibilities) to hold, but only the original model.

So far we've only defined the notion of an 'intervention variable'. But the definition of an intervention in terms of an intervention variable is simple enough. Setting aside some subtleties discussed by Woodward (2003, 94–114), we can think of an intervention on X with respect to Y as simply X's taking some value $X = x$ as a result of an intervention variable $I_{X,Y}$ taking one of its switching values. So, in **Early Preemption,** $I_{M,D}$ taking $I_{M,D} = 1$ serves as an intervention on M with respect to D that sets $M = 0$. The interventionist's proposal is that when evaluating counterfactuals for the purposes of analyzing causation, we should suppose their antecedents are realised by interventions. This is taken to be equivalent to adopting the 'equation replacement' method of evaluating counterfactuals introduced in Section 4.6.1. That's because both methods involve a change to the target variable, X, that preserves the functional relationship between X and Y as well as affecting *neither* the values of ancestors of Y that don't lie on a pathway between X and Y *nor* any influence that those ancestors have on Y that isn't mediated by X.[54] In the next section, we'll see that the notion of an intervention can also potentially help with analysing probabilistic causation.[55]

5 Probabilistic Causation

David Hume – whose regularity theory was examined in Section 2 – wrote at a time when Newtonian mechanics ruled the roost. While it has recently been argued that Newtonian mechanics isn't strictly deterministic,[56] it can be treated as such for most intents and purposes, and in any case doesn't yield well-defined non-trivial (i.e non-0 or 1) probabilities for the unfolding of events over time. At a time when it was reasonable to think that the universe was fundamentally deterministic, it might have seemed plausible to think that all causes are, in the circumstances in which they occur, lawfully sufficient for their effects.

But Newtonian mechanics is now known not to be strictly correct, and has been supplanted by quantum mechanics (QM) and GTR. On the orthodox

---

[54] The assumption that it's possible to intervene upon each variable V in a model of a causal system – thus overriding the usual functional dependence described by the equation for V – without impacting the functional relationships described by any of the other equations in the model is referred to as the assumption of 'modularity'. For a defence of this assumption, see Hausman and Woodward (1999); for criticism, see Cartwright (2002, 2004).

[55] Note that, on the foregoing characterisation of an intervention, interventions needn't be human actions. Rather, natural processes can serve as interventions provided they have the formal characteristics described in this section (see Woodward 2003, 103–4). So there's nothing particularly anthropocentric about adopting an interventionist semantics for the counterfactuals needed to analyse causation.

[56] See Earman (1986, esp. 29–40), Malament (2008, 803–4) and Norton (2008).

interpretation of QM, and on several heterodox interpretations, the world is fundamentally probabilistic. If one of these interpretations is correct, then there might be causes that are neither sufficient for their effects nor elements of sets of conditions that suffice for their effects. Rather, at least some causes might (in the circumstances in which they occur) simply confer a certain probability upon their effects. The example **Probabilistic Bomb**, given in Section 2.5.2 is an example of causation involving quantum probabilities and was seen to pose (insurmountable) difficulties for the regularity approach.

Cases of probabilistic causation don't just pose problems for regularity analyses, but also for the counterfactual analyses examined in the previous section. To see this, suppose that, in **Probabilistic Bomb**, the bomb is itself an inherently unstable device: it has some probability of spontaneously exploding even if the threshold reading of the Geiger isn't reached. Perhaps the bomb is a nuclear device with a Uranium-235 core. There's some (small) probability that enough U-235 atoms undergo spontaneous α-decay to start a chain reaction resulting in the bomb's exploding. In such a case, it appears to be *false* that if I hadn't placed the radioactive material near the Geiger, the bomb wouldn't have exploded. After all, it might still have (spontaneously) exploded, since there would have been a residual probability of its doing so. [57] Indeed it might still have exploded at more-or-less the same time at which it did in fact explode (and perhaps – depending on the workings of the bomb – in much the same manner).[58]

Probabilistic Bomb might seem a contrived example: it's a deliberate attempt to show that quantum probabilities can have significant macroscopic consequences. But it's not just in fundamental physics that probabilities figure heavily. Probabilities are also widespread throughout the higher-level sciences, including statistical mechanics, genetics, ecology, chemistry, meteorology, and economics, to name just a few. There are some deep and difficult scientific

---

[57] As Lewis (1986b, 180–4) argues, because the probabilities involved here are (assuming orthodox QM) objective, it's simply wrong to think that, although the bomb would still have had a certain (small) probability of exploding, there's nevertheless some fact of the matter about whether it *would* or *wouldn't* have exploded in such circumstances.

[58] One might object as follows (thanks to an anonymous referee for pressing me on this point). Suppose that, for example, the bomb's detonation mechanism consists in the shooting of one piece of U-235 into another and that in the actual world the impact between the two pieces occurs at 12 noon. Then, one might argue, if an insignificant number of U-235 atoms in the bomb's core in fact spontaneously decayed prior to 12 noon, then it's true after all that, if I hadn't placed my radioactive material near the Geiger, then the bomb wouldn't have exploded. A worry about this response, however, is that there would still have been a chance of significant spontaneous decay occurring *just after* 12 noon. One may be forced to accept reasonably stringent standards of event fragility if one is to maintain that the resultant explosion is a different event from that which occurred in the actual world and therefore that the explosion counterfactually depended upon my action.

and philosophical questions to be asked concerning (a) how, if at all, the probabilities of these higher-level sciences relate to the probabilities of QM; and (potentially relatedly) (b) to what extent these probabilities are objective, as opposed to simply being introduced into scientific models and theories to reflect our partial ignorance of the world.[59] What seems clear however is that, firstly, viewed through a macro-lens our world appears to be probabilistic and, secondly, in light of the fact that probabilistic theories prevail at every level all the way down to QM, it's doubtful whether, even if we knew enough of the fine details of the universe, we'd be able to uncover sufficient (or indeed necessary) causes for each effect.

To illustrate, consider the causal relation between asbestos inhalation and mesothelioma. The process by which the former brings about the latter is complex, and there isn't yet consensus on the details. But one study (Yang et al. 2010) suggests the following. Asbestos is toxic to human mesothelial cells (which line the pleural cavities in the lungs). The damage that asbestos fibres do to these cells results in the protein HMGB1 being released from them. The presence of extracellular HMGB1 leads macrophages (a type of white blood cell) to release another protein, TNF-α. TNF-α activates NF-κB in damaged mesothelial cells. Activated NF-κB helps some of these damaged cells to survive. These damaged cells can then develop into cancer cells.

Assume that some picture like this is right. Then, even in the case of an individual person, the causal process by which asbestos exposure can bring about mesothelioma is probabilistic. It depends, for instance, upon the 'random walks' followed by asbestos fibres experiencing Brownian motion in the air and in the lungs. It depends upon the probabilistic process of osmosis (of HMGB1 and TNF-α), which comes under the purview of statistical mechanics. And it depends upon molecular bonding processes (which are involved in the stimulation of macrophages by HMGB1, and in the activation of NF-κB by TNF-α), which are subject to QM probabilities.

It thus seems that we ought to make room in our theorising about causation for probabilistic causation since it appears that at least some, and perhaps even the vast majority, of causation in our universe is probabilistic. Many scholars working in the regularity and counterfactual traditions explicitly acknowledge that their accounts are designed only to handle the deterministic case (see, e.g., Lewis 2004a, 79–80; Baumgartner 2013, 87; Strevens 2007, 95; and Hitchcock 2001a, 275; 2007, 498). In what follows, we'll explore some of the leading accounts of probabilistic causation. In the course of doing so, we'll see that the

---

[59] These questions are discussed by, inter alia, Albert (2000), Loewer (2001), and Callender and Cohen (2010).

challenges confronting accounts of deterministic causation – such as preemption – have their probabilistic analogues, but that there are also novel challenges that arise in the probabilistic case.

5.1 Accounts of Probabilistic Causation

A number of authors have proposed accounts of probabilistic causation that appeal to a notion of *probability-raising*. Classic accounts include those of Suppes (1970), Reichenbach (1971, see esp. 204), and Lewis (1986b, Postscript B). In **Probabilistic Bomb**, for example, my placing the radioactive material near to the Geiger raises the probability of the bomb's exploding, even though the bomb may have retained some residual probability of exploding (due to its inherent instability) had I not done so.

There's more than one way of understanding the notion of 'probability-raising'. The two most popular are in terms of (i) conditional probabilities; and (ii) counterfactuals. The latter approach can thus be seen as an extension, to the probabilistic case, of the counterfactual tradition of attempts to analyse causation. But first let's focus on the conditional probability approach.[60] Let's also initially focus upon actual causation (of which **Probabilistic Bomb** is an instance).[61]

Let c and e be token events with $O(c)$ and $O(e)$ being, respectively, the proposition that c occurs and the proposition that e occurs. Then, in the conditional probability sense, c's raising the probability of e is understood in terms of the following inequality:

$$P(O(e)|O(c)) > P(O(e)|\neg O(c)) \qquad (1)$$

This says that the probability of e's occurring conditional upon c's occurring is greater than the probability of e's occurring conditional upon c's not occurring: in **Probabilistic Bomb**, for example, the probability of the bomb's exploding conditional upon my placing the radioactive material near the Geiger is greater than the probability of the bomb's exploding conditional upon my not doing so.

There are, however, at least three problems with understanding probability-raising in terms of conditional probabilities if causation is, in turn, to be understood in terms of probability-raising (see Lewis 1986b, 178–9). The first

---

[60] This is the more traditional approach, adopted by Suppes (1970) and Reichenbach (1971), among others.

[61] Probabilistic type-causation is discussed in Section 5.6.

is that it's a straightforward consequence of the probability calculus that if Inequality 1 holds, then so does:

$$P(O(c)|O(e)) > P(O(c)|\neg O(e)) \tag{2}$$

Understood in terms of conditional probabilities, probability-raising therefore doesn't have the asymmetry that we associate with causation.

A second, related problem is that, understood in the conditional probability sense, independent effects of a common cause typically raise the probability of one another. Recalling the example of Section 2.5.3, the probability of a storm is higher conditional upon a barometer predicting a storm than it is conditional upon the barometer's not doing so. The barometer would be useless if its readings were not correlated with the weather in this way. But this correlation doesn't exist because the barometer reading is a cause of the storm, but rather because an earlier fall in atmospheric pressure is very probable *conditional* upon the barometer's prediction of a storm and a storm is very probable *conditional* upon a fall in atmospheric pressure.

One way of responding to these two problems is as follows. First, it appears that the probabilities of an event evolve over time, at least when we consider objective probabilities (Lewis 1986b, 91) of the sort that are presumably relevant to analysing causation.[62] For instance, in **Probabilistic Bomb**, the probability of the bomb's exploding once the Geiger threshold is reached is higher than prior to its being reached. So we need some specification of *which time's* probabilities are the relevant ones for evaluating causal claims. A natural suggestion is that we should consider the (conditional) probabilities that obtain just before the putative cause occurs.

Now suppose that c is in fact a cause of e. Then, assuming that causes occur prior to their effects, there should be a time t sufficiently close to e's occurrence by which c has already occurred. Once it has occurred, c has a probability 1 of occurring. So, at t, c has a probability 1 of occurring. It follows via the probability calculus that, at t, the probability of c occurring conditional *either* upon the occurrence of e *or* upon the non-occurrence of e is also 1 (provided e doesn't have probability 1 by that time – an issue we'll discuss shortly). But then e *doesn't* raise the probability of c according to the time t probabilities, which are the relevant ones for assessing whether e is a cause of c.[63] This approach

---

[62] I'll say more about the interpretation of probabilities in Section 5.6.

[63] A residual concern is that there might not be a time prior to the occurrence of e by which all of e's causes have occurred. For instance, if time is continuous and there's no unmediated action at a temporal distance, then presumably e must have an infinite series of more and more temporally proximate causes. So, no matter what time prior to the occurrence of e we choose, some of

also helps with the problem of common causes. If, say, we take the probabilities relevant to assessing whether the barometer reading was a cause of the storm to be those that obtain immediately before the barometer indicates a storm, then since the atmospheric pressure has already fallen by that time (and so has probability 1 of doing so), the barometer reading by then doesn't raise the probability of the atmospheric pressure falling and hence doesn't raise the probability of the storm.[64]

There is, however, a third problem with understanding probability-raising in terms of conditional probabilities. This problem is more technical. Conditional probabilities are standardly defined as a ratio:

$$P(A|B) = \frac{P(A\&B)}{P(B)} \qquad (3)$$

But ratios are undefined when the denominator is equal to zero. Now suppose that event c has probability 1 of occurring shortly before it does occur. Then, at that time, $P(\neg O(c)) = 0$ and so the conditional probability on the RHS of Inequality 1 is undefined. The concern is thus that, if probability-raising is understood in terms of conditional probabilities, we appear not to be able to say that events that have probability 1 (shortly before their occurrence) raise the probability of (and therefore cause) anything.

It's possible that, in our world, no event has probability 1 before – even shortly before – it occurs. But even if so, wouldn't we like our probabilistic analysis of causation to apply to deterministic worlds too? A plausible thought

---

e's causes won't yet have occurred by that time and might retain a probability less than 1. If so, we're liable to get the result that e raises the probability of *those* causes in the conditional probability sense.

[64] The solution under consideration in the last two paragraphs of the main text has appealed to the notion that objective chance is time-relative. Certainly we often speak as though this is so. For instance, as I write, it's reasonable to say 'the chance of average global temperatures increasing by > 0.5°C by 2050 is increasing each day the international community puts off tough decisions'. But we can allow this without committing to the notion that time-relativity is a *fundamental* feature of chance. For example, Hoefer (2007, 562) regards chances as relativised to chance setups (e.g. a particular coin flip might be a chance setup relative to which the chance of *heads* is 0.5) and not as in any fundamental way relativised to times (Hoefer 2007, sect. 3.2). Commenting on David Lewis's view of chances as time-relative, Hoefer (2007, 564–5) says that '[f]or Lewis, a non-trivial time-indexed objective probability $Pr_t(A)$ is, in effect, the chance of A occurring given the instantiation of a big setup: the entire history of the world up to time t.' (Hoefer himself makes the plausible allowance that objective chances may be defined given setups that are more limited: for instance, there's a well-defined chance of a silver atom's being deflected in a particular direction given that it's fired through a Stern-Gerlach device.) One way of construing this is to adopt an analysis of time-relativised objective probabilities according to which $P_t(A) =_{def} P(A|H_t)$ where H_t is a proposition characterising the entire history of the world through time t. That is, objective probability isn't inherently time-relativised, but a time-relativised notion of objective probability can be defined by conditioning the objective probability distribution upon history up to the time in question.

is that deterministic worlds are simply the special case where the set of each event's causes raises its probability all the way to 1. But in a deterministic world each event has probability 1 of occurring (shortly before it occurs). So it seems that no event could raise the probability of any other in the conditional probability sense. Since there surely is causation in deterministic worlds (if Bohmian mechanics turns out to be true then *ours* is a deterministic world – at least at the fundamental level), a probabilistic analysis that invoked conditional probabilities would fail to apply to some possible cases of causation.[65]

The counterfactual approach to probability-raising, advocated by Lewis (1986b, 175–9), seemingly overcomes these difficulties. To see how the counterfactual approach interprets probability-raising, suppose that c and e are actual events and that, at a time t just *after* the occurrence of c, the (unconditional) probability of e's occurring is $P_t(O(e)) = x$ (the subscript to the probability function indicates that this is the objective probability distribution that obtains at t). Then consider the following counterfactual:

$$\neg O(c) \;\square\!\!\rightarrow\; P_t(O(e)) < x \qquad\qquad (4)$$

This says that, if c hadn't happened, then the probability at time t of e occurring would have less than x (the actual probability of e at time t). This is a counterfactual understanding of what it is for c to raise e's probability.

This approach has the capacity to overcome the problems associated with the conditional probability approach (Lewis 1986b, 175–9). For, firstly, the fact that $\neg O(c)$ has zero probability in the actual world doesn't imply that there's no world in which it's true, and doesn't stand in the way of $O(e)$ having a well-defined probability in such a world. For instance, in a deterministic world, it's plausible that, where c and e are actual events and c is a non-overdetermining cause of e, then $P(\neg O(c)) = 0$ and $P(O(e)) = 1$ and yet, evaluated with respect to such a world, it's a true counterfactual that if c hadn't occurred, then the probability of e would have been 0.

Second, assuming counterfactuals don't backtrack, this approach avoids the problem of effects raising the probability of their causes and also the problem of common causes. That's because, on the non-backtracking reading of counterfactuals, it's not true that (for example) if the barometer hadn't said there was going to be a storm, then the probability of the atmospheric pressure having earlier fallen would have been lower. Rather, because the atmospheric pressure has already fallen, its probability of doing so is 1 by a time t just after

---

[65] At best it would seem that it could serve as an *empirical analysis* of causation, in the sense discussed in Section 3.4.1.

the barometer says there's going to be a storm, and it's also 1 at t in the nearest worlds (according to a similarity metric that implies non-backtracking) in which the barometer doesn't indicate that there's going to be a storm. Moreover, because the barometer reading doesn't raise the probability of the fall in atmospheric pressure in the counterfactual sense, it doesn't raise the probability of the storm either.

As in the deterministic context, one can achieve the desired non-backtracking results about counterfactuals by appealing to interventions (as an alternative to Lewis's similarity semantics). Though it will be easier to be formally precise about the notion of an 'intervention' in the probabilistic case once we have the notion of a probabilistic causal model – to be introduced in Section 5.3 – in hand, we can for the time being just think of interventions as exogenous manipulations of variable values. As in the deterministic case, variables can be used to represent the occurrence or non-occurrence of events.

Goldszmidt and Pearl (1992) introduce a special notation $do(X = x)$ to represent that the value of variable X is set to $X = x$ by means of an intervention. They then use $P(\cdot|do(X = x))$ to represent the probability distribution that would result from setting X to $X = x$ by means of an intervention. Using that notation, the interventionist counterfactual approach to probability-raising says that – where C and E are binary variables which each have the possible values 1 and 0 – $C = 1$ raises the probability of $E = 1$ iff:

$$P(E = 1|do(C = 1)) > P(E = 1|do(C = 0)) \qquad (5)$$

Despite the resemblance in the notation, terms like $P(E = 1|do(C = 0))$ don't represent conditional probabilities. Rather, they represent counterfactual probabilities: the probability for $E = 1$ that *would* obtain if C were intervened upon to set $C = 0$. Thus Inequality 5 says that the probability for $E = 1$ that *would* obtain if C were intervened upon to set $C = 1$ is higher than the probability of $E = 1$ that would obtain if C were intervened upon to set $C = 0$.[66] In virtue of the non-backtracking character of interventionist counterfactuals, this inequality won't obtain if the events represented by $C = 1$ and $E = 1$ are independent

---

[66] Given that $C = 1$ must be the actual value of C if (the state represented by) $C = 1$ is to be an actual cause of (the state represented by) $E = 1$, couldn't we simply have the actual unconditional probability $P(E = 1)$ that obtains shortly after the occurrence of the state represented by $C = 1$ figure on the LHS of Inequality 5 (as opposed to the probability for $E = 1$ that would have obtained under an intervention setting $C = 1$)? The answer is 'yes' if one is considering whether $C = 1$ straightforwardly raises the probability of $E = 1$. However, as we'll see in the next subsection, some approaches to probabilistic causation appeal to a notion of latent probability-raising that's revealed by holding certain other variables fixed (by interventions). On some approaches we're allowed to hold common causes of C and E fixed at non-actual values, in which case an intervention may well be needed to return C to its actual value.

effects of a common cause or if $E = 1$ is the cause and $C = 1$ is the effect. For instance, an intervention on the barometer reading with respect to the atmospheric pressure or to the storm might consist of an experimenter grabbing the barometer needle and twisting it so that it no longer points at the word 'storm' and instead points at 'fair'. But, in such circumstances, the probability of the fall in atmospheric pressure or of the storm wouldn't be any lower.

5.2 Difficulties

Even when understood in a sense that rules out probability-raising between independent effects of a common cause and probability-raising of causes by their effects, there are (insurmountable) difficulties for any account that seeks simply to identify causation with probability-raising. Cases of probabilistic preemption can be used to illustrate that (even when understood in such a sense) probability-raising is neither necessary nor sufficient for causation (Menzies 1989, 1996). An example is provided by the following variant on **Probabilistic Bomb**.

Probabilistic Preemption

Someone (neither you nor I) has connected a Geiger counter to a bomb so that the bomb will explode if the Geiger registers above a threshold reading. I place a chunk of U-232, which has a half-life of 68.9 years and decays by α-emission, near the Geiger. By chance, enough U-232 atoms decay within a short enough interval for the Geiger to reach the threshold reading so that the bomb explodes. Unbeknownst to me, you've been standing nearby observing. You have a chunk of Th-228, which has a half-life of 1.9 years and decays by α-emission, which contains at least as many atoms as my chunk of U-232. You've decided that you'll place your Th-228 near the Geiger iff I fail to place my U-232 there. Seeing that I place my U-232 near the Geiger, you don't place your Th-228 there.

In this case, my placing my U-232 near the Geiger is plausibly regarded as a cause of the bomb's exploding. Let M, D, Y, T, and E be binary variables which, respectively, take value 1 if the following things occur (and 0 otherwise): I place my U-232 near the Geiger; you decide to place your Th-228 near the Geiger iff I don't place my U-232 near the Geiger; you place your Th-228 near the Geiger; the threshold Geiger reading is reached; the bomb explodes. Though my act ($M = 1$) is a cause of the explosion ($E = 1$), the following inequality holds:

$$P(E = 1|do(M = 1)) < P(E = 1|do(M = 0)) \qquad (6)$$

That is, my placing my U-232 near the Geiger *lowers* the probability of the bomb's exploding. That's because it strongly lowers the probability of your placing your more potent Th-228 near the Geiger (your Th-228 is more potent because Th-228 has a shorter half life than U-232 and your chunk of Th-228 contains at least as many atoms as my chunk of U-232). Probability-raising (even in the interventionist counterfactual sense) is therefore unnecessary for causation.

Probabilistic Preemption also illustrates the fact that probability-raising (even in the interventionist counterfactual sense) isn't sufficient for causation. Specifically, notice that your decision ($D = 1$) to place your Th-228 near the Geiger iff I don't place my U-232 there raises the probability of the explosion.[67] That is:

$$P(E = 1|do(D = 1)) > P(E = 1|do(D = 0))   \qquad (7)$$

Specifically, Inequality 7 holds provided there's some chance that $M = 0$, because $D = 1$ raises the probability that the bomb will still explode in the scenario in which $M = 0$. Yet your decision ($D = 1$) *isn't* a cause of the explosion since you don't place your Th-228 near the Geiger.[68]

Recall that one promising response to the fact that *counterfactual dependence* isn't necessary for causation in the deterministic case was to appeal to a notion of de facto dependence, which we formalised with the help of SEMs. It turns out that there's an analogous response to the fact that *probability-raising* isn't necessary for causation: namely to appeal to a notion of de facto probability-raising, that is, latent probability-raising that's revealed by holding fixed certain features of the actual situation.[69] In the present example,

---

[67] At least if we assume that the alternative is for you to decide not to place your Th-228 near the Geiger come what may. More will be said about causal contrastivity in the probabilistic case in Section 5.4.

[68] Cases of probability-raising without causation are discussed by Menzies (1989), Schaffer (2001), and Hitchcock (2004), inter alia. Cases of causation without probability-raising are discussed by Hesslow (1976), Rosen (1978), and Salmon (1980), inter alia.

[69] Such an approach is taken by Hitchcock (2001b), Kvart (2004), and Fenton-Glynn (2017), inter alia. Cartwright (1979, 423) endorses an an analogous approach to understanding type-causation: specifically that a type-level factor C is a cause of a type-level factor E iff 'C increases the probability of E in every [population] which is otherwise causally homogenous with respect to E', where a situation is 'causally homogenous' with respect to E iff all other causal factors for E are held fixed. Note that in the case of type causation it generally makes no sense to speak of holding factors fixed 'at their actual values', since the values will vary across populations. So Cartwright effectively requires that C increases the probability of E no matter what values we hold these other factors fixed at. For a critique of this requirement, see Carroll (1991). Eells (1991, 86–7) offers a similar definition of what he calls 'positive causal relevance', but also defines 'negative causal relevance' (when C *lowers* the probability of E no matter what values we hold these other factors fixed at), and 'mixed causal relevance' (e.g. when C raises the probability of E for some assignments of values to these other factors, but fails to raise it for others).

although my placing my U-232 near the Geiger doesn't raise the probability of the bomb's exploding, latent probability-raising is revealed by holding fixed the fact that you don't place your Th-228 there. On the interventionist approach, 'holding fixed' your not placing your Th-228 there means intervening on the variable Y to keep it at $Y = 0$ whilst varying the value of M by means of a further intervention. The result, captured by Inequality 8, is that latent probability-raising is revealed:

$$P(E = 1|do(Y = 0 \& M = 1)) > P(E = 1|do(Y = 0 \& M = 0)) \quad (8)$$

This inequality says that the probability of the explosion if interventions are made to ensure that you don't place your radioactive material near the Geiger but I place mine there is higher than the probability of the explosion if interventions are made to ensure that neither of us places our radioactive material near the Geiger.

Just as the features of the actual situation to be held fixed in seeking de facto dependence in the deterministic case could be more clearly identified with the aid of SEMs, so the features of the actual situation to be held fixed in seeking de facto probability-raising can be more clearly identified with the help of probabilistic causal models. Though the converse problem of probability-raising without causation doesn't have such a clear analogue in the deterministic case – it's much more plausible to take counterfactual dependence to be sufficient for causation than it is to take probability-raising to be sufficient – it turns out that appeal to probabilistic causal models opens up possible avenues for dealing with the problem of probability-raising without causation too.

5.3 Probabilistic Causal Models

Adopting the interventionist framework discussed in the previous subsection, a probabilistic causal model $\mathcal{M}^*$,[70] can be thought of as comprising a set of variables, $\mathcal{V}$, together with a function, $do(\cdot)$. The $do(\cdot)$ function takes as input a value of a variable in $\mathcal{V} - do(X_1 = x_1)$ – or a conjunction of values of different variables in $\mathcal{V} - do(X_1 = x_1 \& \dots \& X_n = x_n)$ – and returns a probability distribution over the possible values of the variables in $\mathcal{V}$: the probability distribution that would obtain if $X_1 \& \dots \& X_n$ were set to the values $X_1 = x_1 \& \dots \& X_n = x_n$ by interventions.

We'll return to the question of whether (latent) probability-raising rather than, say, probability-lowering should – as many accounts seem to assume – play a special role in our causal thinking in Section 5.7.

[70] I'll use asterisks to indicate that a model is a probabilistic model, rather than a SEM.

We can construct a graphical representation of a probabilistic model by taking the variables in $\mathcal{V}$ as the nodes of the graph and drawing a directed edge ('arrow') from a variable X_i to a variable X_j $(X_i, X_j \in \mathcal{V})$ just in case there's some assignment of values to the other variables in $\mathcal{V}$ such that the value of V_i makes a difference to the probability distribution over the values of V_j when the other variables in $\mathcal{V}$ are held fixed by interventions at the assigned values.

To illustrate, consider **Probabilistic Preemption**. The key features of the $do(\cdot)$ function – to be explored in more detail in the following – can be inferred from our statement of **Probabilistic Preemption**, whilst the remainder of our model – which we might call $\mathcal{PP}^*$ – is summarised as follows.

$\mathcal{PP}^*$

Variables: $\{M, D, Y, T, E\}$

Interpretation: $M = 1/0$ if I do/don't place my U-232 near the Geiger; $D = 1/0$ if you do/don't decide to place your Th-228 near the Geiger *iff I don't place my U-232 near the Geiger*; $Y = 1/0$ if you do/don't place your Th-228 near the Geiger; $T = 1/0$ if the threshold Geiger reading is/isn't reached; $E = 1/0$ if the bomb does/doesn't explode.

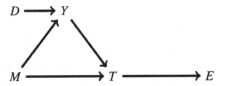

Figure 6: Probabilistic Preemption

As an illustration of our rule for drawing arrows in a causal graph, consider the arrow from M to T in Figure 6 (the graphical representation of $\mathcal{PP}^*$). This arrow exists because there's an assignment of values to all of the other variables in our model – one such assignment is $\{D = 0, Y = 0, E = 0\}$ – such that if those variables are held fixed at the assigned values by interventions, then intervening to change the value of M makes a difference to the probability distribution over the values of T. That is:

$$P(T = 1|do(D = 0 \& Y = 0 \& E = 0 \& M = 1))$$
$$\neq P(T = 1|do(D = 0 \& Y = 0 \& E = 0 \& M = 0)) \qquad (9)$$

That is, if interventions occur to ensure that (i) you don't make the decision that you do; (ii) you don't place your radioactive material near the Geiger;

and (iii) the bomb doesn't explode, then interventions changing whether or not I place my radioactive material near the Geiger make a difference to the probability that the threshold is reached.[71] It's key here that an intervention to prevent the bomb exploding would be something – such as the severing of the wire between the Geiger and the bomb – that doesn't affect whether the threshold is reached. This is needed to secure the relevant non-backtracking reading of the counterfactuals. The justification for the inclusion of each of the other arrows in Figure 6 is analogous to the justification for the inclusion of that from M to T.

Now consider why there isn't, for instance, an arrow from Y to E. That's because there's no assignment of values to all other variables in the model such that, when the variables are held fixed at those values by interventions, intervening on Y makes a difference to the probability distribution over E. In particular, any such assignment includes an assignment of a value to T but, when the value of T is held fixed by an intervention (either at 1 or 0), then the value of Y no longer has a bearing on the probability that $E = 1$. Next, consider why there isn't an arrow from D to M. This is simply because your decision is completely irrelevant to my action and can't be made probabilistically relevant to it by holding fixed by interventions the other variables in the model.[72] Finally, consider why there isn't an arrow from E to T. This is because, as mentioned in the previous paragraph, an intervention on E would only count as such if it doesn't affect T. When the value of E is changed in such a way, its value

---

[71] It's important to say something about which time's probabilities are in question. One suggestion (assumed in what follows) is that, where X_i represents an event that occurs at t then, in evaluating whether there's an arrow from X_i to X_j, the probabilities to consider are those that would obtain just after t if X_i had been intervened on *and* interventions had already occurred by t to fix the values of all variables in the model besides X_j. The fact that some of those other variables represent events that occur later than t doesn't stand in the way of interventions fixing their values having occurred prior to t. For example, the wire connecting the Geiger and the bomb could already have been severed to ensure that the bomb doesn't explode (an intervention on the value of E) by the time I place my radioactive material near the Geiger.

One might wonder whether, in a probabilistic world, it's reasonable to suppose that an intervention occurring by t could deterministically fix the value of a variable E representing an event occurring sometime later. In fact, it's not strictly necessary to appeal to deterministic interventions to get the results described in the main text. For instance, an intervention that (merely) sufficiently lowered the probability of your placing your radioactive material near the Geiger (even in the circumstance in which I don't place mine there) would ensure that my action raises the probability that the threshold is reached. Still, the complication of probabilistic interventions isn't an avenue we'll pursue further here.

[72] In the language of graph theory, Y acts as an 'unshielded collider' for D and M (Spirtes et al. 2000, 10): that is, there are arrows from each of D and M to Y but there's no arrow connecting D and M. While probabilistically independent variables can be rendered dependent by *conditioning* upon an unshielded collider, they can't be rendered dependent by *intervening* on an unshielded collider.

has no bearing upon that of T (no matter what values we hold the other variables in the model fixed at). Each of the other arrow absences in Figure 6 has a justification analogous to the justification for the absence of an arrow from Y to E, from D to M, or from E to T.

Just as, in seeking to analyse deterministic causation in terms of SEMs appeal to an 'apt' SEM is needed, so too in seeking an analysis of (probabilistic) causation in terms of probabilistic models, the notion of an 'apt' probabilistic model is needed. The requirements (Partition), (Independence), (Naturalness), and (Serious Possibilities) placed on SEMs carry across to probabilistic models without need of modification (and with the same motivation). (Veridicality) however needs some modification, or at least elaboration, to make clear what it means for a probabilistic model to 'entail only true counterfactuals'. The requirement (Veridicality*) is what we need to substitute in its place.

(Veridicality*) For any conjunction $X_1 = x_1 \& \ldots \& X_n = x_n$[73] (where $X_1 \ldots X_n$ are variables in the model and $x_1 \ldots x_n$ are possible values of those variables) taken as an input, the probability distribution $P(\cdot | do(X_1 = x_1 \& \ldots \& X_n = x_n))$ yielded as output by the $do(\cdot)$ function of the model should be the objective probability distribution over the variables in the model that would result from interventions setting $X_1 = x_1 \& \ldots \& X_n = x_n$.[74]

The requirement (Veridicality*) is thus the requirement that the model (specifically its $do(\cdot)$ function) entail true counterfactuals about what the objective probabilities would be under all combinations of values of the variables in the model's variable set. The models that I consider in what follows all satisfy (Veridicality*). Specifically, in describing various causal scenarios, I'll stipulate what the objective probabilities would be under various counterfactual suppositions about their variables. I shall then only describe models of these scenarios that entail the correct values for these counterfactual chances.

The only requirement left to discuss now is (Stability). This can be applied to probabilistic models in exactly the same form (and with the same motivation) as it was applied to SEMs in Section 4.6.3, except that its reference to (Veridicality) should be replaced by a reference to (Veridicality*).

---

[73] The simple expression $X_1 = x_1$ shall be treated as a limiting case of such a conjunction.

[74] In light of the discussion of Footnote 71, we should strictly speaking require that, where the actual values of $X_1 \ldots X_n$ represent states obtaining at $t_1 \ldots t_n$ the $do(\cdot)$ function entails, for each t_i, the objective probability distribution over the variables in the model that would result had interventions setting $X_1 = x_1 \& \ldots \& X_n = x_n$ occurred by t_i.

5.4 Analysing Causation in terms of Probabilistic Models

With the notion of an apt probabilistic model in hand, we're in a position to describe an analysis of (actual) causation that captures the notion of de facto probability-raising more precisely. The following is a first-pass analysis that will be of heuristic value. By analysing causation in terms of de facto probability-raising it helps with the problem, described in Section 5.2, of causation without probability-raising. But, as we'll see, it can't be our final analysis because it doesn't help with the converse difficulty of probability-raising without causation.

PC

Where x and x' $(x \neq x')$ are possible values of the variable X and y is a possible value of the variable Y, then X taking the value $X = x$ rather than $X = x'$ is a cause of Y taking the value $Y = y$ iff $X = x$ and $Y = y$ are the actual values of X and Y and, in some apt probabilistic model $\mathcal{M}^*$, there's a path $\mathcal{P}$ from X to Y such that – where $\mathbf{W}$ is the set of variables that lie off the path $\mathcal{P}$ and the actual values of those variables are $\mathbf{W} = \mathbf{w}^*$ – $\mathcal{M}^*$ entails that:

$$P(Y = y | do(\mathbf{W} = \mathbf{w}^* \& X = x)) > P(Y = y | do(\mathbf{W} = \mathbf{w}^* \& X = x'))$$

(IN)

Analysis **PC** is highly analogous to **SEA**.[75] **SEA** instructs us to look for a path $\mathcal{P}$ in an apt SEM such that holding fixed the variables that lie off $\mathcal{P}$ at their actual values reveals a relation of de facto counterfactual dependence of $Y = y$ upon $X = x$. Analogously, **PC** instructs us to look for a path $\mathcal{P}$ in an apt probabilistic model such that holding fixed the variables that lie off $\mathcal{P}$ at their actual values reveals a relation of de facto probability-raising of $Y = y$ by $X = x$.

Analysis **PC** counts $M = 1$ (rather than $M = 0$) as a cause of $E = 1$ in $\mathcal{PP}^*$. To see this, consider the path $\langle M, T, E \rangle$. There are two variables – D and Y – that lie off this path. Moreover, holding D and Y fixed at their actual values

---

[75] One *disanalogy* is that, while **SEA** builds in contrastivity on both the cause and the effect sides, **PC** builds in contrastivity on the cause side only. Though it's possible to incorporate contrastivity on the effect side too, this is a bit less natural in the probabilistic case where there's often not a determinate fact about what value the effect variable *would* have taken (but only a probability distribution over alternative values) had the cause variable taken its contrast value.

$D = 1$ and $Y = 0$ reveals a relation of de facto probability-raising of $E = 1$ by $M = 1$, as required by condition **IN** of **PC**:

$$P(E=1|do(D=1\&Y=0\&M=1)) > P(E=1|do(D=1\&Y=0\&M=0)) \tag{10}$$

This simply says that, given your decision but your failure to place your material near the Geiger, the probability of the explosion is higher if I place my material there than if I don't. This is of course true in **Probabilistic Preemption**.

Analysis **PC** is, however, in danger of giving the incorrect result that your decision to place your material near the Geiger iff I don't place mine there, $D = 1$, is also a cause of the explosion, $E = 1$. To see this, first note that it's presumably the case that if *both* of us place our material near the Geiger then the probability of the threshold being reached is higher than if I alone place my material there. Next note that, since the case is probabilistic, we might reasonably allow that your decision merely makes it very likely that you won't place your radioactive material near the Geiger if I do rather than deterministically preventing you from doing so. But now consider the path $\langle D, Y, E \rangle$ from D to E. There's one variable – M – that lies off this path. But it could be that $D = 1$ raises the probability of $E = 1$ even holding M fixed at its actual value $M = 1$.[76]

$$P(E = 1|do(M = 1\&D = 1)) > P(E = 1|do(M = 1\&D = 0)) \tag{11}$$

That is, holding fixed that I place my material near the Geiger, your decision still raises the probability of the bomb's exploding because there's some chance given your decision that you'll place your material near the Geiger too. In this case, **PC** counts your decision as a cause even though you don't in fact place your material near the Geiger.

To think about how to modify **PC** to give a more adequate analysis of causation, it helps to consider *why* we don't regard your decision as a cause. The key reason is presumably that your placing your material near the Geiger (something which you don't in fact do) is an essential part of the causal process via which your decision had the potential to bring about the explosion. Of

---

[76] To secure this result we need to assume that the alternative to your making the decision that you do is for you to make some decision (e.g. to simply walk away from the scene) that makes it less likely than it in fact is that you will *also* place your material near the Geiger. Now we could either simply stipulate this, or we could model the case with a multi-valued variable representing the various possible decisions open to you. We could then make the explicitly contrastive point that **PC** (incorrectly) treats your making the decision that you do *rather than (e.g.) deciding to walk away* as a cause of the bomb's explosion.

course, it wouldn't be very helpful to simply build into our analysis the require-ment that 'there be a complete causal process' from putative cause to putative effect without further characterisation of what a 'causal process' is and what it is for such a process to be incomplete.[77] And, as discussed in Section 1, it seems impossible to give an account of causal processes without falling back on counterfactuals (which, in the probabilistic case, might include counterfactuals about probabilities).

Fortunately, we can sidestep these issues because it appears there's a probabi-listic symptom of the incompleteness of a causal process. Take the relationship between your decision and the bomb's explosion and note that, although the probability of the bomb's exploding may (even holding fixed my placing my radioactive material near the Geiger) be higher if you take your decision than if you don't, the probability of the bomb's exploding given that you take your decision *but you don't place your radioactive material near the Geiger* is (even holding fixed my placing my material near the Geiger) no higher than if you'd never taken your decision in the first place. That is, although Inequality 11 holds, it's also the case that:

$$P(E=1|do(M=1\&D=1\&Y=0)) \leq P(E=1|do(M=1\&D=0)) \quad (12)$$

There's thus a sense in which the (de facto) probability-raising of $E=1$ by $D=1$ is 'non-robust' in that it doesn't hold up when we take into account the actual value of a variable on the path from D to E.

By contrast, the de facto probability-raising of $E=1$ by $M=1$ revealed by Inequality 10 *is* robust. That is, it continues to hold when we take into account the actual value of the only intermediate variable, T, on the path $\langle M, T, E \rangle$ along which we discovered de facto probability-raising. That is:

$$P(E=1|do(D=1\&Y=0\&M=1\&T=1))$$
$$> P(E=1|do(D=1\&Y=0\&M=0)) \quad (13)$$

In words, holding fixed the fact that you make your decision but don't place your material near the Geiger, the probability of the bomb's exploding is (not only higher if I place my material there than if I don't, but is) higher if I place my material there *and the threshold is reached* than if I don't place my material there in the first place.

---

[77] Though, as we saw in Section 3.4.3, an account needn't be reductive to be illuminating, we would surely like an account that tells us something about what a causal process *is* rather than simply falling back on intuitions about this notion.

This suggests that we might try to revise **PC** so that the revised analysis requires not merely that there's de facto probability-raising of effect by cause, but also that this probability-raising be suitably robust. This is the approach taken by Fenton-Glynn (2017). The following is a slightly streamlined version of that analysis.

PC1

Where x and x' ($x \neq x'$) are possible values of the variable X and y is a possible value of the variable Y, then X taking the value $X = x$ rather than $X = x'$ is a cause of Y taking the value $Y = y$ iff $X = x$ and $Y = y$ are the actual values of X and Y and, in some apt probabilistic model $\mathcal{M}^*$, there's a path $\mathcal{P}$ from X to Y such that – where $\mathbf{W}$ is the set of variables that lie off $\mathcal{P}$, where $\mathbf{Z}$ is the set of variables that lie intermediate between X and Y on $\mathcal{P}$, and where the actual values of the variables in $\mathbf{W}$ are $\mathbf{W} = \mathbf{w}^*$ – then for all subsets $\mathbf{Z'}$ of $\mathbf{Z}$ (where $\mathbf{Z'} = \mathbf{z}^*$ are the actual values of the variables in $\mathbf{Z'}$), $\mathcal{M}^*$ entails that:

$$P(Y=y|do(\mathbf{W}=\mathbf{w}^*\&X=x\&\mathbf{Z'}=\mathbf{z}^*)) > P(Y=y|do(\mathbf{W}=\mathbf{w}^*\&X=x'))$$

(IN1)

Analysis **PC1** is complex and one can best get to grips with it by seeing it in action. To start with, observe that it yields the correct results about **Probabilistic Preemption**. First, it correctly counts $M = 1$ as a cause of $E = 1$. To see this, consider the path $\langle M, T, E \rangle$ in $\mathcal{PP}^*$. There are two variables – D and Y – that lie off this path, and one – T – that lies intermediate between M and E upon it. The set of intermediate on-path variables is therefore just $\{T\}$ which has two subsets: $\{T\}$ and $\emptyset$. Inequality 10 shows that $M = 1$ raises the probability of $E = 1$ when we hold the off-path variables fixed at their actual values. This de facto probability-raising is trivially robust (in the sense captured by Inequality **IN1** of **PC1**) relative to the empty subset of $\{T\}$. But it's also robust relative to the non-empty subset of $\{T\}$. This is indicated by the holding of Inequality 13. The fact that the de facto probability-raising of $E = 1$ by $M = 1$ is robust relative to both subsets of the set of intermediate on-path variables, $\{T\}$, is precisely what is required for **PC1** to diagnose $M = 1$ as a cause of $E = 1$.

Second, **PC1** yields the correct result that $D = 1$ isn't a cause of $E = 1$. To see this note that $\langle D, Y, T, E \rangle$ is the only path from D to E in $\mathcal{PP}^*$. The variable M is the only one that lies off this path and its actual value is $M = 1$. Moreover, although $D = 1$ may raise the probability of $E = 1$ when we hold fixed that $M = 1$, as revealed by the holding of Inequality 11, this probability-raising isn't robust relative to the subset $\{Y\}$ of the intermediate on-path variables, as is revealed by the holding of Inequality 12.

Probabilistic Preemption is an early preemption case. But it's quite easy to verify that **PC1** yields the correct results in certain cases of late preemption too. Consider Figure 3 from Section 4.6.2. This was a graph of the SEM that we constructed for **Late Preemption**. But Figure 3 could just as well represent a version of **Late Preemption** which differs from the original in that each of the relations is merely probabilistic: specifically, Bob's intention merely makes it likely that Bob will shoot if the deer isn't dead by 1:00pm, Alice's shot merely makes it likely that the deer will be dead by 1:00pm, Bob's shooting would merely make it likely that the deer would be dead by 1:01pm if Alice didn't shoot. In such a case, **PC1** entails that Alice's shot $A = 1$ was a cause of $D_{1:01} = 1$. To see this, consider the path $\langle A, D_{1:00}, D_{1:01} \rangle$. $A = 1$ raises the probability of $D_{1:01} = 1$ when we hold the two off-path variables – BI and B – fixed at their actual values $BI = 1$ and $B = 0$. That is:

$$P(D_{1:01} = 1 | do(BI = 1 \& B = 0 \& A = 1))$$
$$> P(D_{1:01} = 1 | do(BI = 1 \& B = 0 \& A = 0)) \qquad (14)$$

That is, given that Bob intends to shoot if the deer isn't dead by 1:00pm, but in fact Bob doesn't shoot, the probability that the deer is dead by 1:01pm is higher if Alice shoots than if she doesn't.

This de facto probability-raising is robust in the sense that it continues to hold when we take into account the actual value of the (only) intermediate on-path variable $D_{1:00}$, namely $D_{1:00} = 1$. That is:

$$P(D_{1:01} = 1 | do(BI = 1 \& B = 0 \& A = 1 \& D_{1:00} = 1))$$
$$> P(D_{1:01} = 1 | do(BI = 1 \& B = 0 \& A = 0)) \qquad (15)$$

This says that, holding fixed that Bob intends to shoot if the deer isn't dead by 1:00pm but in fact Bob doesn't shoot, the probability that the deer is dead by 1:01pm is higher if Alice shoots *and* the deer is dead by 1:00pm than if simply Alice doesn't shoot in the first place.

Analysis **PC1** also implies that Alice's shooting was a cause of the deer's being dead by 1:00pm. To see this, consider the path $\langle A, D_{1:00} \rangle$. Holding fixed the off-path variables BI, B, and $D_{1:01}$ at their actual values $BI = 1$, $B = 0$, and $D_{1:01} = .1$, $A = 1$ raises the probability of $D_{1:00} = 1$. That is:

$$P(D_{1:00} = 1 | do(BI = 1 \& B = 0 \& D_{1:01} = 1 \& A = 1))$$
$$> P(D_{1:00} = 1 | do(BI = 1 \& B = 0 \& D_{1:01} = 1 \& A = 0)) \qquad (16)$$

Note that an intervention to ensure that $D_{1:01} = 1$ wouldn't count as such if it worked by ensuring (or even affecting the probability) that $D_{1:00} = 1$, which is why we can be sure that Inequality 16 holds. Since there are no variables intermediate between A and $D_{1:00}$ on the path $\langle A, D_{1:00} \rangle$ it's automatically true that the de facto probability-raising revealed by Inequality 16 is suitably robust.[78] Hence **PC1** correctly counts $A = 1$ as a cause of $D_{1:00} = 1$.

Analysis **PC1** also yields the correct result that Bob's intention wasn't a cause of the deer's being dead by 1:01pm (it implies that it wasn't a cause of the deer's being dead by 1:00pm simply in virtue of the absence of a path from BI to $D_{1:00}$[79]). To see this observe that there's only one path – $\langle BI, B, D_{1:01} \rangle$ – from BI to $D_{1:01}$ but that $BI = 1$ doesn't raise the probability of $D_{1:01}$ when we hold fixed the off-path variables A and $D_{1:00}$ at their actual values, $A = 1$ and $D_{1:00} = 1$. Specifically:

$$P(D_{1:01} = 1 | do(A = 1 \& D_{1:00} = 1 \& BI = 1)) = 1$$
$$= P(D_{1:01} = 1 | do(A = 1 \& D_{1:00} = 1 \& BI = 0)) \qquad (17)$$

This might seem a bit of a 'cheat' because it relies on the fact that the connection between the variables $D_{1:00}$ and $D_{1:01}$ is deterministic: given that the deer is dead at 1:00pm it's dead at 1:01pm with probability 1. But note that, even if this link were merely probabilistic (perhaps there's some probability of the deer's being resuscitated between 1:00pm and 1:01pm!), **PC1** would still yield the correct result. Specifically, in that case, although $BI = 1$ would de facto raise the probability of $D_{1:01} = 1$ (Equality 17 would – assuming that, given Bob's intention, there's some chance of his shooting even if the deer is dead at 1:00pm – now be converted into an inequality, with the term on the LHS being greater than the term on the RHS), this de facto probability-raising wouldn't be robust once we took account of the actual value – $B = 0$ – of the intermediate on-path variable in the usual way. That is:

$$P(D_{1:01} = 1 | do(A = 1 \& D_{1:00} = 1 \& BI = 1 \& B = 0))$$
$$= P(D_{1:01} = 1 | do(A = 1 \& D_{1:00} = 1 \& BI = 0)) \qquad (18)$$

---

[78] Note that, since the causal process connecting Alice's shot to the deer's being dead at 1:00pm is complete, interpolating variables along the path $\langle A, D_{1:00} \rangle$ wouldn't affect the fact that this probability-raising is robust. For instance, (holding the off-path variables fixed at their actual values) the probability of the deer's being dead at 1:00pm would be higher if Alice shot *and* her speeding bullet passed mid-air between her and the deer than it would be if Alice hadn't shot in the first place. So interpolating a variable M representing the presence of the speeding bullet along the path $\langle A, D_{1:00} \rangle$ wouldn't affect the verdict of **PC1**.

[79] Plausibly, the absence of such a path is no mere idiosyncrasy of our model, but rather will be a feature of any model of the scenario satisfying (Veridicality*). That's because the absence of such a path reflects the assumption of the example that Bob's intention is to give Alice enough of an opportunity to kill the deer that he won't himself shoot the deer dead by 1:00pm.

That is (even holding the off-path variables fixed at their actual values) the probability of the deer's being dead by 1:01pm given Bob's intention but also the fact that Bob doesn't shoot is no higher than it would be if simply Bob hadn't had the intention in the first place.

5.5 Overdetermination of Probabilities

It's at least in principle possible that probabilistic scenarios analogous to **Symmetric Overdetermination** should arise. These are cases in which an effect isn't overdetermined but, as it were, its probability is overdetermined. The following is an example.[80]

Overdetermination of Probabilities

Consider a specific one-second interval, i, and suppose the probability that atom a will decay within i is (i) 0.1 if a doesn't absorb a neutron at the start of i; (ii) 0.8 if a absorbs neutron b alone or neutron c alone at the start of i; and (iii) 0.8 if a absorbs *both* neutron b *and* neutron c at the start of i. Suppose, moreover, that the chance of a absorbing both b and c at the start of i is 1 and there's probability 0 that a will absorb any other particle during i. In fact, a absorbs b and c simultaneously at the start of i and decays within i.

Suppose that, although absorbing neutrons brings about decay *probabilistically*, there's no intermediate *mechanism* via which it brings about decays. That is, absorbing a neutron results in a certain probability of decay and then either the decay occurs or it doesn't, with no further mechanism needed to bring about the decay.

I don't claim that this example is physically realistic, just that a case with this probabilistic structure is conceivable and, insofar as we'd like our analysis to apply across possible worlds, then our analysis ought to give a reasonable treatment of cases like this. Moreover, symmetric overdetermination can be considered the limiting case of overdetermination of probabilities where the probability is overdetermined to be 1. And presumably we'd like our analysis to be able to handle cases of symmetric overdetermination if we have ambitions for it to hold in deterministic worlds.

Consider a model $\mathcal{OP}^*$ of **Overdetermination of Probabilities** with a $do(\cdot)$ function that entails the probabilities specified in **Overdetermination of Probabilities**, and with the additional features described as follows.

---

[80] I stipulate that some of the probabilities in this example are 0 or 1 for simplicity. More complex examples can be given in which all probabilities are between 0 and 1.

$\mathcal{OP}^*$

Variables: $\{BA, CA, AD\}$

Interpretation: $BA = 1/0$ if b is/isn't absorbed by a at the beginning of **i**; $CA = 1/0$ if c is/isn't absorbed by a at the beginning of **i**; $AD = 1/0$ if a does/doesn't decay within **i**.

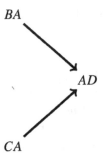

Figure 7: Overdetermination of Probabilities

Relative to $\mathcal{OP}^*$, **PC1** counts *neither* $BA = 1$ *nor* $CA = 1$ as a cause of $AD = 1$ (which is analogous to the way in which **SEA** doesn't count overdeterminers as causes). To see this in the case of $BA = 1$ (the reasoning is exactly analogous for $CA = 1$) note that there's only one path from $BA = 1$ to $AD = 1$: namely $\langle BA, AD \rangle$. There's one variable lying off this path: namely CA. But $BA = 1$ doesn't raise the probability of $AD = 1$ when we hold fixed the off-path variable at its actual value, $CA = 1$. Specifically:

$$P(AD=1|do(CA=1\&BA=1))=0.8=P(AD=1|do(CA=1\ BA=0))\ (19)$$

I suspect that, as in cases of symmetric overdetermination so in cases of symmetric overdetermination of probabilities more generally, the majority of people will have the opinion that both 'overdeterminers' are causes, or at least that this is a preferable result to counting neither as causes. This, then, is a limitation of **PC1** (and it's easily verified that it similarly fails to count symmetric overdeterminers as causes in true overdetermination cases). One might therefore be tempted to consider a liberalisation of **PC1** that allows that we may sometimes *vary* features of the actual situation in seeking latent probability-raising of effect by the putative cause. This would be analogous to the liberalisation of **SEA** considered in Section 4.6.4. For instance, were we allowed to suppose that $CA = 0$, then we'd recover probability-raising of $AD = 1$ by $BA = 1$ in **Overdetermination of Probabilities**.

5.6 Type Causation, Interventions, and Epistemology

The focus of this section has been on actual causation: the sort of token causal relation that's usually at issue when scientists, lawyers, and pretty much everyone else talks about causation between particulars rather than types. But, just as in the deterministic case, we might sometimes have use for a notion of causation between variables themselves (as discussed in Section 4.6.5) and not just between variable values.

In fact, the definitions of relations of causation between variables given by Woodward (2003) and discussed in Section 4.6.5 carry across neatly to the probabilistic case. As in that case, X can be defined as as a *direct cause* of Y in a model iff X is a parent of Y in the model. The only difference from the deterministic case is that the 'model' in question is now understood as a probabilistic model rather than a SEM.

Woodward (2003, 51, 57) himself provides probabilistic versions of his definitions of *total cause* and *contributing cause*. Specifically, X is defined as a (probabilistic) *total cause* of Y iff there's a pair of values x, x' ($x \neq x'$) of X such that, had (due to an intervention) X taken $X = x$, then the probability distribution over the possible values of Y would have been different than had (due to an intervention) X taken $X = x'$. On the other hand, X is defined as a (probabilistic) *contributing cause* of Y relative to a model iff there's a path $\mathcal{P}$ from X to Y in the model, a pair of values x, x' ($x \neq x'$) of X, and an assignment of values to all off-path variables such that, had (due to an intervention) X taken $X = x$ and had (due to interventions) the off-path variables taken the assigned values, then the probability distribution over the possible values of Y would have been different than had (due to an intervention) X taken $X = x'$ and had (due to interventions) the off-path variables taken the assigned values.

With the concept of a probabilistic model together with these definitions of (probabilistic) causation between variables in hand, the notion of an 'intervention' in the probabilistic case can be formally defined by strict analogy to the definition for the deterministic case.[81] Specifically, an intervention on X with respect to Y consists in X's taking some value $X = x$ as a result of an intervention variable $I_{X,Y}$ taking one of its 'switching' values ('switching' is defined in the same way as in Section 4.6.6, but with 'depends' now replaced by 'probabilistically depends'), with $I_{X,Y}$ counting as an intervention variable for X with respect to Y iff (a) $I_{X,Y}$ is a contributing cause of X, (b) $I_{X,Y}$ acts as a switch for X, (c) there's no directed path from $I_{X,Y}$ to Y that doesn't go via X; (d) there's no variable that's an ancestor of $I_{X,Y}$ that lies on a directed path to Y that doesn't run via $I_{X,Y}$.

---

[81] Though, as noted in Footnote 71, in the probabilistic case we might also have use for the notion of a probablistic intervention.

Just as in the deterministic case, the various definitions of causal relations between variables don't necessarily capture type-causal notions. For example, according to these definitions, M counts as a direct, total, and contributing cause of Y in our model $\mathcal{PP}^*$ of **Probabilistic Preemption**. Yet these variables represent whether or not certain token events occur. But, just as in the deterministic case, it's of course possible to use variables to represent things that are non-particular. For instance, we could construct a type-level probabilistic model of the causes of heart disease with variables representing such things as average number of cigarettes smoked, minutes of exercise taken, saturated fat consumed, etc. per day. Such a model would be 'type-level' in the sense that it can be taken to model the causes of heart disease across a population rather than for any specific individual.

An interesting thing to note about Woodward's definitions of causation between variables is that they don't focus on *probability-raising*, but rather simply require that interventions on the putative cause variable X *make a difference* to the probability distribution over the values of the putative effect variable Y (when other variables are held fixed). Part of the reason for this is that it simply doesn't make sense to talk about X raising the probability of Y where X and Y are variables. What we *can* sensibly say is that the value of X affects the probability distribution over Y. Still, we might ask, when it comes to actual causation (which has been the principal focus of this section) why focus (as we have) on (de facto) probability-*raising* rather than probability-*lowering*, given that both notions are well-defined when we're concerned with the relations between specific variable values? This is an important question and one that will be taken up in the next subsection.

Before turning to that however it is worth observing that, when it comes to probabilistic models, the interpretation of the probabilities figuring in a model of a particular (token) system is presumably different from the interpretation of the probabilities figuring in a 'type-level' model. In giving a probabilistic model of a particular system (an individual person or, for instance, the Geiger counter set-up described in **Probabilistic Preemption**), we're assuming that it makes sense to assign probabilities to particulars: these are sometimes known as 'single-case' probabilities. Such probabilities can't be straightforwardly taken to be frequencies/statistical probabilities. That's because there's only one of the token system in question, so the 'frequencies' (i.e. statistics) for it are trivial. For instance, the bomb either explodes or it doesn't, so the frequency with which it does is either 1/1 (=1) or 0/1 (=0).

Of course, one might take population-level frequencies to imply single-case probabilities. For instance if, on average, Jo Doe smokes 10 cigarettes, exercises for 30 minutes, and consumes 20g of saturated fat per day, then one might take her individual probability of suffering heart disease to equal the population-level frequency with which people who on average smoke 10-a-day, exercise for 30 minutes, and consume 20g of saturated fat develop heart disease. Thus a type-level model might be used to calculate the risk of heart disease in a particular person. But this is far from an innocuous move, since it gives rise to the notorious *reference-class* problem.[82]

While this isn't the place to enter into extended discussion of the issue, it's worth pointing out that there are alternative ways of understanding single-case probabilities, including 'propensity' interpretations (e.g. Popper 1959; Hacking 1965; Mellor 1971) and 'best system' interpretations (e.g. Lewis 1994). When it comes to 'type-level' models, on the other hand, a frequency interpretation is tempting, but it's not the only possibility. For instance, the 'best system' approach provides an interpretation of type-level as well as single-case probabilities.

When we're dealing with models that represent type-level phenomena and frequency/statistical information about them, there are quite sophisticated techniques for causal discovery (see, e.g., Spirtes et al. 2000). In particular, we can often discover a lot about the causal structure (e.g. direct, total, and contributing causal relations among variables) simply by examining the conditional and unconditional correlations between variables, at least if we're prepared to make certain general assumptions about the relationship between causation and probability. Such general assumptions include that events that don't cause one another are probabilistically independent conditional upon any common causes they might have, that if A causes C (only) by causing B then A and C are probabilistically independent conditional upon B,[83] and that independent causes of a single effect become probabilistically dependent when that effect is conditioned upon.[84] We can get even further if we have statistics available from randomised control trials (RCTs). That's because RCTs act like type-level 'interventions': in an RCT, values of some variable X are assigned randomly to members of a population thus, within that population, breaking the usual dependence of X

---

[82] Venn (1866) was the first to systematically describe this problem. Hájek (2007) provides a contemporary discussion.

[83] These two assumptions are captured by the so-called Causal Markov Condition (Spirtes et al. 2000, 11).

[84] To illustrate: patients with MS are often misdiagnosed as having lupus and vice versa because many of the symptoms are the same. Loss of coordination, for example, is an effect of both conditions. So consider the class of patients presenting with coordination loss. We would expect that the incidence of MS is lower in that subset of such patients who have lupus than it is among that subset of such patients who don't have lupus. This explains why, for instance, a doctor presented with such a patient who rules out lupus becomes more confident that MS is the correct diagnosis and, conversely, why a doctor who discovers that such a patient has lupus wouldn't usually investigate whether the patient *additionally* has MS.

upon its parents (i.e. acting as a 'switch'). This allows us to determine the causal influence of X upon a target variable Y while being reasonably confident that any association between X and Y isn't due to common causes or indeed due to Y's being a cause of X.

It's unquestionable that our discovery – by such methods – of type-level causal relations and statistical information about their strength informs our judgements about causal relations and probabilities at the token level. But the relationship between the type- and the token level, as ever, isn't straightforward. For instance, statistical associations at the type level needn't be reflected by genuine indeterminism at the token level (still less need they match the single-case probabilities): probabilities at the type level may reflect unmodelled causes, measurement errors and the like. Nor can actual causal relations simply be 'read off' from those at the type level. For instance, driving with a damaged tyre might be a type-level cause of car crashes (because of the risk of blow out) but, in a particular case where a driver drives with a damaged tyre and crashes, the former needn't be an actual cause of the latter (suppose the driver fell asleep at the wheel and there was no blow-out). However, delving deeper into the complexities surrounding the relation between type and token levels is a task that must await another occasion.

5.7 Probability-Raising vs Probability-Lowering

In the previous subsection, we deferred an important question. This is the question of why, when we are analysing actual causation, we focus on (de facto) probability-*raising* rather than probability-*lowering*? To address this question, it is helpful to consider an argument given by Mellor (1995, 58–66, ch. 7) who (as we saw in Section 3.4.1) claims that causation has several important 'connotations', including that effects are evidence for their causes, that causes explain their effects, and that causes are means for bringing about their effects as ends. Mellor says:

> Any apparent cause that failed to raise its effects' chances would not only be neither evidence for them nor explain them, it would not be a means to them if they were ends. And nothing that satisfied none of these three connotations would be a cause: the whole point of calling it a cause would be lost. In particular, to call something a cause that provides no way of bringing about its effects seems to me an obvious contradiction in terms. (Mellor 1995, 88)

Anyone who wishes to allow that there can be probability-lowering causes – as my placing my U-232 near the Geiger appeared to be in **Probabilistic Preemption** – has either to reject the notion that causation has the connotations that Mellor describes, or has to reject the idea that probability-raising is necessary to satisfy these connotations. Doing the former seems rather unpalatable, but the advocate of the de facto probability-raising approach can explain why doing the latter is less so.

Continuing to focus on the example of **Probabilistic Preemption**, note that, given that you didn't actually place your Th-228 near the Geiger in **Probabilistic Preemption**, my placing my U-232 there clearly *explains* why the bomb exploded. Also fairly clear is that my placing my U-232 near the Geiger was a *means* to the bomb's exploding. Of course, there's a sense in which my failure to place my U-232 there would *also* have been a means (indeed a more effective one) to the bomb's exploding. Yet, in the scenario, I didn't know this since I didn't know of your presence. Nevertheless, although it's plausibly a platitude about causation that causes are means to their effects, it's surely not a platitude that for anything to count as a cause it must be the *best* means to its effect or even the best available in the circumstances. For example, if someone who works in asbestos removal also smokes, surely her smoking has the potential to cause her to suffer lung cancer even if she could have more surely brought about the latter by removing her protective mask at work.[85]

Is my placing my U-232 *evidence* that the bomb exploded? It depends. What is evidence for what presumably depends upon the epistemic state of the explainee. For instance, it would be odd to say that the causes of the explosition count as evidence of its occurrence for someone who actually witnessed the explosion. Likewise, for someone who hasn't witnessed the explosion but who knows of your intention to place your Th-228 near to the Geiger if I don't place my U-232 there, the fact that I *do* place my U-232 there is plausibly evidence *against* the explosion. Still, for someone who doesn't know of your intention, or who knows of your intention but also knows that in fact you don't place your Th-228 near to the Geiger, my action is presumably evidence that the bomb explodes. I take it that this is sufficient for the relevant connotation of causation to be satisfied.

Interestingly, then, if the de facto probability-raising account is correct, then it might be true to say that c caused e (since c de facto raised the probability of

---

[85] Of course there's a disanalogy between this case and **Probabilistic Preemption** in that smoking and not wearing a protective mask to work are not mutually exclusive, but my placing my U-232 near the Geiger and my not doing so are. So perhaps one could argue that, where c is a member of a set S of mutually exclusive (and jointly exhaustive?) possibilities, c is only a cause of e if c is the best means to e in S? Yet I suspect that, in doing so, one leaves the realm of straightforward 'connotations' or platitudes about causation for the realm of philosophical theorising. Moreover, such an argument seems to simply beg the question against the advocate of the de facto probability-raising account.

e) even though there's a sense in which *e* occurred *despite c* (since *c* straight-forwardly lowered the probability of *e*).[86] But arguably this is just as it should be: in **Probabilistic Preemption** we appear to be tempted to regard my action as a cause, even though there's also a sense in which the explosion occurred 'despite' my action. Focusing on the role that my action played in preventing your more efficacious action can incline us to adopt the 'despite' locution. The advocate of the de facto probability-raising approach will therefore argue that there's no contradiction between asserting '*c* caused *e*' and that '*e* occurred despite *c*', though perhaps – because the former emphasises *c*'s role in bring-ing about *e*, while the latter emphasises *c*'s (unmanifested) potential to prevent *e* – it's infelicitous to utter them in the same breath.

This leaves the question of why our concept of actual causation should track de facto probability-raising rather than overall probability-raising? Presumably the advocate of the de facto probability-raising account will seek to answer this in terms of the connotations of (actual) causation: for instance, the fact that my placing my U-232 near the Geiger would serve as evidence and explanation for the explosion for an agent who knew enough about the circumstances to know that you didn't place your Th-228 there, and that I successfully performed my action as a means to the explosion. In contrast, consider the following example due to Cartwright (1979, 425):

> I consider eradicating the poison oak at the bottom of my garden by spraying it with defoliant. The can of defoliant claims that the spray is 90 per cent effective; that is, the probability of a plant's dying given that it is sprayed is .9, and that the probability of its surviving is .1.

Spraying the plant lowers the probability of the plant's survival. But if the plant survives then, as Cartwright points out, the spraying can't explain its survival. Likewise the spraying is neither evidence for the survival nor is it in any sense a means to bringing about the survival. Plausibly this is because the spraying doesn't even bear a de facto probability-raising relation to the survival, and for this reason isn't an actual cause.

6 Conclusion

In this Element, we've examined three broad traditions in the philosophy of causation: the regularity, counterfactual, and probabilistic approaches. Our discussion showed how contemporary thinking about causation has been influ-enced by a long history of intellectual thought on the subject, but we've also

---

[86] Cf. Hitchcock (2001b, esp. 382).

ventured to the cutting edge of contemporary research: for instance, by examining structural equation and probabilistic modelling approaches. The latter approaches constitute a rich – and increasingly interdisciplinary – vein of current investigation. In addition to examining particular approaches, we've investigated issues that are sometimes informed by, but to some extent transcend, particular theories of causation: for instance, questions about the relationship between type and token causation, about causation involving absences, and about what kinds of, and how many, relata the causal relation should be taken to have.

The discussion has, of necessity, been selective. For reasons discussed in the Introduction, we haven't examined the rich literatures on causal processes and causal mechanisms. There are also some general questions about causation that we haven't tackled, such as whether causation comes in degrees (Halpern and Hitchcock 2015, Bernstein 2017, Sartorio 2020) and whether Hart and Honoré (1959) and Lewis (1973a) are indeed correct that our tendency to pick out abnormal factors as causes is something to be explained by a pragmatics of causal discourse, as opposed to a more recent suggestion that it's to be baked into the semantics of causal claims.[87] Yet by introducing the reader to core theories, concepts, and tools present in the contemporary debate it is hoped, not merely that the reader's enthusiasm for further investigation has been piqued, but also that they have been equipped to get the most out of such investigation, and indeed to participate in the debate themselves.

Of the approaches to causation that we've examined, some appear more promising than others. For my part, I find approaches that appeal to de facto dependence/probability-raising to be more promising than some of the earlier accounts in the counterfactual and probabilistic traditions. Most investigators these days seem to regard these traditions as more fruitful than the regularity tradition, though I suspect that ways of sophisticating the regularity approach (e.g. Strevens 2007, Baumgartner 2013) have been unfairly neglected. Yet even among the approaches that I've suggested are most promising, the specific accounts that have been developed – while elegantly dealing with some of the problems that afflicted earlier accounts – are not immune to difficulties.

Of course the hope is that further progress can be made – these research programmes are very much alive and kicking – but if anyone has come up with an entirely problem- and controversy-free account of *any* philosophically interesting notion, I've yet to see it! Part of the difficulty is that the various aspects of our pre-theoretical thinking about notions such as causation probably

---

[87] For debate over this latter point, see e.g. (McGrath 2005), (Halpern and Hitchcock 2015) and (Blanchard and Schaffer 2015).

don't admit of a complete and elegant regimentation. Yet even if no single, unified account is possible, it's to be hoped that we'll find that we have a range of accounts available each of which captures an important aspect of our causal thinking and whatever it might be 'in the world' that that thinking tracks.

In any case, not all progress in deepening and sharpening our understanding of a phenomenon like causation consists in constructing analyses of it. Much progress consists in getting a clearer picture of the terrain by, for instance, gaining clarity on the nature and variety of difficult cases (e.g. the varieties of preemption, the different ways causation can come apart from probability-raising, the existence of apparent counterexamples to causal transitivity, etc.) and in developing tools and concepts (e.g. sophisticated counterfactual semantics, the notion of an 'intervention', the concept of an ENF counterfactual, the formal techniques of causal modelling and graph theory) for thinking about these and other cases. At the very least, this increased understanding of the nature of causation can help us avoid making basic errors in our thinking about it and, given the centrality of causation to the legal and scientific worlds, the importance of this can't be overstated.

References

Albert, D. Z. (2000). *Time and Chance*. Cambridge, MA: Harvard University Press.

Anscombe, G. E. M. (1971). *Causality and Determination*. Cambridge: Cambridge University Press.

Armstrong, D. M. (1983). *What Is a Law of Nature?* Cambridge: Cambridge University Press.

Baumgartner, M. (2013). A regularity theoretic approach to actual causation. *Erkenntnis 78*, 85–109.

Beebee, H. (2004). Causing and nothingness. In J. Collins, N. Hall, and L.A. Paul (eds): *Causation and Counterfactuals*, pp. 291–308. Cambridge, MA: MIT Press.

Bernstein, S. (2017). Causal proportions and moral responsibility. In D. Shoemaker (ed.), *Oxford Studies in Agency and Responsibility*: vol. 4, pp. 165–82. Oxford: Oxford University Press.

Bigaj, T. (2012). Causation without influence. *Erkenntnis 76*, 1–22.

Blanchard, T. and J. Schaffer (2015). Cause without default. In H. Beebee, C. Hitchcock, and H. Price (Eds.), *Making a Difference*. Oxford: Oxford University Press.

Callender, C. and J. Cohen (2010). Special sciences, conspiracy and the better best system account of lawhood. *Erkenntnis 73*, 427–47.

Carlson, G. and F. Pelletier (Eds.) (1995). *The Generic Book*, Chicago, IL. University of Chicago Press.

Carroll, J. (1991). Property-level causation? *Philosophical Studies 63*, 245–70.

Cartwright, N. (1979). Causal laws and effective strategies. *Noûs 13*, 419–37.

Cartwright, N. (2002). Against modularity, the causal markov condition, and any link between the two: Comments on Hausman and Woodward. *British Journal for the Philosophy of Science 53*, 411–53.

Cartwright, N. (2004). Causation: One word, many things. *Philosophy of Science 71*, 805–19.

Cat, J. (2017). The unity of science, In E. Zalta (ed.): *Stanford Encyclopedia of Philosophy*, https://plato.stanford.edu/archives/fall2017/entries/scientific-unity/

Choi, S. (2002). Causation and gerrymandered world lines: A critique of Salmon. *Philosophy of Science 69*, 105–17.

Collins, J. (2000). Preemptive prevention. *Journal of Philosophy 97*, 223–34.

Craver, C. and J. Tabery (2019). Mechanisms in science. In E. Zalta (ed.): *Stanford Encyclopedia of Philosophy*, https://plato.stanford.edu/archives/sum2019/entries/science-mechanisms/

Davidson, D. (1967). Causal relations. *Journal of Philosophy 64*, 691–703.

Dowe, P. (1992). Wesley Salmon's process theory of causality and the conserved quantity theory. *Philosophy of Science 59*, 195–216.

Dowe, P. (2000). *Physical Causation*. Cambridge: Cambridge University Press.

Dunn, J. (2011). Fried eggs, thermodynamics, and the special sciences. *The British Journal for the Philosophy of Science 62*(1), 71–98.

Eagle, A. (2015). Generic causation. Unpublished Manuscript; http://philpapers.org/rec/EAGGC.

Earman, J. (1986). *A Primer on Determinism*. Dordrecht: Reidel.

Edgington, D. (1997). Mellor on chance and causation. *British Journal for the Philosophy of Science 48*, 411–33.

Eells, E. (1991). *Probabilistic Causality*. Cambridge: Cambridge University Press.

Elga, A. (2001). Statistical mechanics and the asymmetry of counterfactual dependence. *Philosophy of Science 68 (Supplement)*, S313–S324.

Emery, N. (2015). Chance, possibility, and explanation. *British Journal for the Philosophy of Science 66*, 95–120.

Evans, R. (2013). *Altered Pasts: Counterfactuals in History*. Waltham, MA: Brandeis University Press.

Fenton-Glynn, L. (2017). A proposed probabilistic extension of the Halpern and Pearl definition of 'actual cause'. *British Journal for the Philosophy of Science 68*, 1061–124.

Feynman, R. (1965). *The Character of Physical Law*. London: BBC.

Glennan, S. (1996). Mechanisms and the nature of causation. *Erkenntnis 44*, 49–71.

Glennan, S. (2017). *The New Mechanical Philosophy*. Oxford: Oxford University Press.

Glymour, C., D. Danks, B. Glymour et al. (2010). Actual causation: A stone soup essay. *Synthese 175*, 169–92.

Glynn, L. (2013). Of miracles and interventions. *Erkenntnis 78*, 43–64.

Goldszmidt, M. and J. Pearl (1992). Rank-based systems: A simple approach to belief revision, belief update, and reasoning about evidence and actions. In *Proceedings of the Third International Conference on Knowledge Representation and Reasoning*, San Mateo, CA, pp. 661–72. Morgan Kaufmann.

Hacking, I. (1965). *The Logic of Statistical Inference*. Cambridge: Cambridge University Press.

Hájek, A. (2007). The reference class problem is your problem too. *Synthese 156*, 563–85.

Hall, N. (2000). Causation and the price of transitivity. *Journal of Philosophy 97*, 198–222.

Hall, N. (2004). Two concepts of causation. In *Causation and Counterfactuals*, pp. 225–76. Cambridge, MA: MIT Press.

Halpern, J. Y. (2015). Appropriate causal models and stability of causation. Unpublished manuscript, www.cs.cornell.edu/home/halpern/papers/causalmodeling.pdf.

Halpern, J. Y. (2016). *Actual Causality*. Cambridge, MA: MIT Press.

Halpern, J. Y. and C. Hitchcock (2010). Actual causation and the art of modeling. In R. Dechter, H. Geffner, and J. Y. Halpern (Eds.), *Heuristics, Probability and Causality: A Tribute to Judea Pearl*, pp. 383–406. London: College Publications.

Halpern, J. Y. and C. Hitchcock (2015). Graded causation and defaults. *British Journal for the Philosophy of Science 66*, 413–57.

Halpern, J. Y. and J. Pearl (2005). Causes and explanations: A structural-model approach. Part I: Causes. *British Journal for the Philosophy of Science 56*, 843–87.

Hart, H. L. A. and T. Honoré (1959). *Causation in the Law*. Oxford: Clarendon Press.

Hausman, D. and J. Woodward (1999). Independence, invariance and the causal markov condition. *British Journal for the Philosophy of Science 50*, 521–83.

Hesslow, G. (1976). Two notes on the probabilistic approach to causality. *Philosophy of Science 43*, 290–2.

Hitchcock, C. (1995). Salmon on explanatory relevance. *Philosophy of Science 62*, 304–20.

Hitchcock, C. (1996). Farewell to binary causation. *Canadian Journal of Philosophy 26*, 267–82.

Hitchcock, C. (2001a). The intransitivity of causation revealed in equations and graphs. *Journal of Philosophy 98*, 194–202.

Hitchcock, C. (2001b). A tale of two effects. *Philosophical Review 110*, 361–96.

Hitchcock, C. (2004). Do all and only causes raise the probabilities of effects? In J. Collins, N. Hall, and L. Paul (eds.), *Causation and Counterfactuals*, pp. 403–17. Cambridge, MA: MIT Press.

Hitchcock, C. (2007). Prevention, preemption, and the principle of sufficient reason. *Philosophical Review 116*, 495–532.

Hitchcock, C. (2009). Problems for the conserved quantity theory: Counterexamples, circularity, and redundancy. *Monist 92*, 72–93.

Hitchcock, C. and J. Knobe (2009). Cause and norm. *Journal of Philosophy 106*, 587–612.

Hoefer, C. (2007). The third way on objective probability: A sceptic's guide to objective chance. *Mind 116*, 549–96.

Hume, D. (1739). *A Treatise of Human Nature*. London: J. Noon.

Hume, D. (1748). *An Enquiry Concerning Human Understanding*. London: A. Millar.

Ismael, J. (2009). Probability in deterministic physics. *Journal of Philosophy 106*, 89–108.

Kim, J. (1973a). Causation, nomic subsumption, and the concept of event. *Journal of Philosophy 70*, 217–36.

Kim, J. (1973b). Causes and counterfactuals. *Journal of Philosophy 70*, 570–72.

Kment, B. (2006). Counterfactuals and explanation. *Mind 155*, 261–10.

Kroedel, T. (2008). Mental causation as multiple causation. *Philosophical Studies 139*, 125–43.

Kvart, I. (2004). Causation: Probabilistic and counterfactual analyses. In *Causation and Counterfactuals*, pp. 359–86. Cambridge, MA: MIT Press.

Lewis, D. (1973a). Causation. *Journal of Philosophy 70*, 556–67.

Lewis, D. (1973b). *Counterfactuals*. Oxford: Oxford University Press.

Lewis, D. (1979). Counterfactual dependence and time's arrow. *Noûs 13*, 455–76.

Lewis, D. (1986a). Events. In D. Lewis (Ed.), *Philosophical Papers*, vol. 2, pp. 241–269. Oxford: Oxford University Press.

Lewis, D. (1986b). *Philosophical Papers*, vol. 2. Oxford: Oxford University Press.

Lewis, D. (1994). Humean supervenience debugged. *Mind 103*(412), 473–90.

Lewis, D. (2004a). Causation as influence. In J. Collins, N. Hall, and L. Paul (eds.), *Causation and Counterfactuals*, pp. 75–106. Cambridge, MA: MIT Press.

Lewis, D. (2004b). Void and object. In J. Collins, N. Hall, and L. Paul (eds.), *Causation and Counterfactuals*, pp. 277–290. Cambridge, MA: MIT Press.

List, C. and M. Pivato (2015). Emergent chance. *Philosophical Review 124*, 119–152.

Loewer, B. (2001). Determinism and chance. *Studies in History and Philosophy of Science Part B: Studies in History and Philosophy of Modern Physics 32*, 609–20.

Machamer, P., L. Darden, and C. Craver (2000). Thinking about mechanisms. *Philosophy of Science 67*, 1–25.

Mackie, J. (1965). Causes and conditions. *American Philosophical Quarterly 2*, 245–64.

Mackie, J. (1980). *The Cement of the Universe*. Oxford: Oxford University Press.

Malament, D. (2008). Norton's slippery slope. *Philosophy of Science 75*, 799–816.

McDermott, M. (2002). Causation: Influence versus sufficiency. *Journal of Philosophy 99*, 84–101.

McGrath, S. (2005). Causation by omission: A dilemma. *Philosophical Studies 123*, 125–48.

Meek, C. and C. Glymour (1994). Conditioning and intervening. *British Journal for the Philosophy of Science 45*, 1001–21.

Mellor, D. H. (1971). *The Matter of Chance*. Cambridge: Cambridge University Press.

Mellor, H. (1995). *The Facts of Causation*. London: Routledge.

Menzies, P. (1989). Probabilistic causation and causal processes: A critique of Lewis. *Philosophy of Science 56*, 642–63.

Menzies, P. (1996). Probabilistic causation and the pre-emption problem. *Mind 105*, 85–117.

Mill, J. (1843). *A System of Logic, Ratiocinative and Inductive*, Vol. 1. London: John W. Parker.

Moore, M. (2009). *Causation and Responsibility: An Essay in Laws, Morals, and Metaphysics*. Oxford: Oxford University Press.

Northcott, R. (2008). Causation and contrast classes. *Philosophical Studies 139*, 111–23.

Norton, J. (2008). The dome: An unexpectedly simple failure of determinism. *Philosophy of Science 75*, 786–98.

O'Connor, T. (2002). *Persons and Causes: The Metaphysics of Free Will*. Oxford: Oxford University Press.

Paul, L. A. and N. Hall (2013). *Causation: A User's Guide*. Oxford: Oxford University Press.

Pearl, J. (2009). *Causality: Models, Reasoning, and Inference* (2nd ed.). Cambridge: Cambridge University Press.

Popper, K. (1959). The propensity interpretation of probability. *British Journal for the Philosophy of Science 10*, 25–42.

Read, R. (ed.) (2014). *The New Hume Debate: Revised Edition*, Abingdon: Routledge.

Reichenbach, H. (1971). *The Direction of Time*. Mineola, NY: Dover Publications. M. Reichenbach (ed.). First published in 1956.

Reutlinger, A., Schurz, G., Hütteman, A. and Jaag, S. (2019). Ceteris paribus laws, In E. Zalta (ed.): *Stanford Encyclopedia of Philosophy*, https://plato.stanford.edu/archives/win2019/entries/ceteris-paribus/

Rosen, D. (1978). In defense of a probabilistic theory of causality. *Philosophy of Science 45*, 604–13.

Salmon, W. (1994). Causality without counterfactuals. *Philosophy of Science 61*, 297–312.

Salmon, W. (1997). Causality and explanation: A reply to two critiques. *Philosophy of Science 64*, 461–77.

Salmon, W. C. (1980). Probabilistic causality. *Pacific Philosophical Quarterly 61*, 50–74.

Sartorio, C. (2005). Causes as difference-makers. *Philosophical Studies 123*, 71–96.

Sartorio, C. (2020). More of a cause? *Journal of Applied Philosophy 37*, 346–63.

Schaffer, J. (2001). Causes as probability raisers of processes. *Journal of Philosophy 98*, 75–92.

Schaffer, J. (2004). Trumping preemption. In J. Collins, N. Hall, and L. A. Paul (eds.), *Causation and Counterfactuals*, pp. 59–73. Cambridge, MA: MIT Press.

Schaffer, J. (2005). Contrastive causation. *Philosophical Review 114*, 327–58.

Schaffer, J. (2013). Causal contextualisms. In M. Blaauw (ed.), *Contrastivism in Philosophy*, pp. 35–63. New York: Routledge.

Spirtes, P., C. Glymour, and R. Scheines (2000). *Causation, Prediction, and Search* (2nd ed.). Cambridge, MA: MIT Press.

Strevens, M. (2007). Mackie remixed. In J. Campbell, M. O'Rourke, and H. Silverstein (eds.) *Causation and Explanation*, pp. 93–118. Cambridge, MA: MIT Press.

Suppes, P. (1970). *A Probabilistic Theory of Causality, Acta Philosophica Fennica*, vol. 24. Amsterdam: North-Holland.

Tucker, A. (1999). Historiographical counterfactuals and historical contingency. *History and Theory 38*, 264–76.

Venn, J. (1866). *The Logic of Chance*. London and Cambridge: Macmillan and Co.

Werndl, C. (2011). On choosing between deterministic and indeterministic models: Underdetermination and indirect evidence. *Synthese 190*, 2243–65.

Weslake, B. (2020). A partial theory of actual causation. Forthcoming in *British Journal for the Philosophy of Science*.

Williamson, T. (2000). *Knowledge and its Limits*. Oxford: Oxford University Press.

Woodward, J. (2003). *Making Things Happen: A Theory of Causal Explanation*. Oxford: Oxford University Press.

Yablo, S. (2002). De facto dependence. *Journal of Philosophy 99*, 130–148.

Yablo, S. (2004). Advertisement for a sketch of an outline of a prototheory of causation. In J. Collins, N. Hall, and L.A. Paul (eds): *Causation and Counterfactuals*, pp. 119–137. Cambridge, MA: MIT Press.

Yang, H., Z. Rivera, S. Jube et al. (2010). Programmed necrosis induced by asbestos in human mesothelial cells causes high-mobility group box 1 protein release and resultant inflammation. *Proceedings of the National Academy of the Sciences 107*, 12611–16.

Cambridge Elements ≡

Philosophy of Science

Jacob Stegenga
University of Cambridge

Jacob Stegenga is Reader in the Department of History and Philosophy of Science at the University of Cambridge. He has published widely on fundamental topics in reasoning and rationality and philosophical problems in medicine and biology. Prior to joining Cambridge, he taught in the United States and Canada, and he received his PhD from the University of California San par Diego.

About the Series

This series of Elements in Philosophy of Science provides an extensive overview of the themes, topics and debates which constitute the philosophy of science. Distinguished specialists provide an up-to-date summary of the results of current research on their topics, as well as offering their own on those topics and drawing original conclusions.

Cambridge Elements ☰

Philosophy of Science

Elements in the Series

A full series listing is available at: www.cambridge.org/EPSC